" 'Why in the world do you want to go down into the sea?'
is a riddle we are often asked by practical people. George
Mallory was asked why he wanted to climb Mt. Everest,
and his answer serves for us, too. 'Because it is there,' he
said. We are obsessed with the incredible realm of
oceanic life waiting to be known. . . .

" I have recounted how the first goggles led us underwater
in simple and irresistible curiosity, and how that impulse
entangled us in diving physiology and engineering,
which produced the compressed-air lung. Our dives are
now animated by the challenge of oceanography. We
have tried to find the entrance of the great hydrosphere
because we feel that the sea age is soon to come."

from the *Epilogue*

THE SILENT WORLD has been written in English by Captain Cousteau who, although a French citizen and Naval officer attended an American school in his youth and has traveled widely in the United States. (In recent years he has lectured in America and Britain on his underwater experiences.) He was assisted in the preparation of this book by James Dugan, a former correspondent of *Yank*, the Army weekly, who has been associated with Captain Cousteau since World War II.

The handheld color work in Ektochrome is the first ever made in significant depths, using artificial light and scientific color correction.

Captain Jacques Cousteau
with
Frederic Dumas

The Silent World

Futura Publications Limited

An Omega Book

First published in Great Britain in 1952
by Hamish Hamilton Limited
First Omega Edition published in 1975
by Futura Publications Limited

The publishers and the authors wish to express their
gratitude to the staff of the *National Geographic Magazine*
for making possible the inclusion of the colour
photographs in this volume

ISBN 0 8600 77136

Printed and bound in Great Britain by
Redwood Burn Limited, Trowbridge & Esher

Contents

To the men who shared in our work

The
Silent World

Captain Cousteau swims under the sea with his pressurized Rollei-
flex camera making color flash photos such as those shown in the
color section. The photos in this book record man's most intimate and
comprehensive adventure inside the sea. Captain Cousteau provides
a running commentary on the great moments of fifteen years and
five thousand dives by the authors and members of the Undersea
Research Group. Most pictures are from his eight movies made
underwater, some are by photographers accompanying his various
expeditions, a few (such as the shot above) have been rendered from
color stereophotos. When a picture is not credited to a photographer,
it was made by Cousteau, Frédéric Dumas or Philippe Tailliez, mem-
bers of the Undersea Research Group. (Photo by Jean de Wouters
d'Oplinter.)

Chapter One **Menfish**

O NE morning in June, 1943, I went to the railway station at Bandol on the French Riviera and received a wooden case expressed from Paris. In it was a new and promising device, the result of years of struggle and dreams, an automatic compressed-air diving lung conceived by Émile Gagnan and myself. I rushed it to Villa Barry where my diving comrades, Philippe Tailliez and Frédéric Dumas waited. No children ever opened a Christmas present with more excitement than ours when we unpacked the first "aqualung." If it worked, diving could be revolutionized.

We found an assembly of three moderate-sized cylinders of compressed air, linked to an air regulator the size of an alarm clock. From the regulator there extended two tubes, joining on a mouthpiece. With this equipment harnessed to the back, a watertight glass mask over the eyes and nose, and rubber foot fins, we intended to make unencumbered flights in the depths of the sea.

We hurried to a sheltered cove which would conceal our activity from curious bathers and Italian occupation troops. I checked the air pressure. The bottles contained air condensed to one hundred and fifty times atmospheric pressure. It was difficult to contain my excitement and discuss calmly the plan of the first dive. Dumas, the best goggle diver in France, would stay on shore keeping warm and rested, ready to dive to my aid, if necessary. My wife, Simone, would swim out on the surface with a schnorkel breathing tube and watch me through her submerged mask. If she signaled anything had gone wrong, Dumas could dive to me in seconds. "Didi," as he was known on the Riviera, could skin dive to sixty feet.

My friends harnessed the three-cylinder block on my back with the regulator riding at the nape of my neck and the hoses looped over my head. I spat on the inside of my shatterproof glass mask and rinsed it in the surf, so that mist would not form inside. I molded the soft rubber flanges of the mask tightly over forehead and cheekbones. I fitted the mouthpiece under my lips and gripped the nodules between my teeth. A vent the size of a paper clip was to pass my inhalations and exhalations beneath the sea. Staggering under the fifty-pound apparatus, I walked with a Charlie Chaplin waddle into the sea.

The diving lung was designed to be slightly buoyant. I reclined in the chilly water to estimate my compliance with Archimedes' principle that a solid body immersed in liquid is buoyed up by a force equal to the weight of the liquid displaced. Dumas justified me with Archimedes by attaching seven pounds of lead to my belt. I sank gently to the sand. I breathed sweet effortless air. There was a faint whistle when I inhaled and a light rippling sound of bubbles when I breathed out. The regulator was adjusting pressure precisely to my needs.

I looked into the sea with the same sense of trespass that I have felt on every dive. A modest canyon opened below, full of dark green weeds, black sea urchins and small flowerlike white algae. Fingerlings browsed in the scene. The sand sloped down into a clear blue infinity. The sun struck so brightly I had to squint. My arms hanging at my sides, I kicked the fins languidly and traveled down, gaining speed, watching the beach reeling past. I stopped kicking and the momentum carried me on a fabulous glide. When I stopped, I slowly emptied my lungs and held my breath. The diminished volume of my body decreased the lifting force of water, and I sank dreamily down. I inhaled a great chestful and retained it. I rose toward the surface.

My human lungs had a new role to play, that of a sensitive ballasting system. I took normal breaths in a slow rhythm, bowed my head and swam smoothly down to thirty feet. I felt no increasing water pressure, which at that depth is twice that of the surface. The aqualung automatically fed me increased compressed air to meet the new pressure layer. Through the fragile human lung linings this counter-pressure was being transmitted to the blood stream and instantly spread throughout the incompressible body. My brain received no subjective news of the pressure. I was at ease, except for a pain in the middle ear and sinus cavities. I swallowed as one does in a landing airplane to open my Eustachian tubes and healed the pain. (I did not wear ear plugs, a dangerous practice when under water. Ear plugs would have trapped a pocket of air between them and the eardrums. Pressure building up in the Eustachian tubes would have forced my eardrums outward, eventually to the bursting point.)

I reached the bottom in a state of transport. A school of silvery sars (goat bream), round and flat as saucers, swam in a rocky chaos. I looked up and saw the surface shining like a defective mirror. In the center of the looking glass was the trim silhouette of Simone, reduced to a doll. I waved. The doll waved at me.

I became fascinated with my exhalations. The bubbles swelled on the way up through lighter pressure layers, but were peculiarly flattened like mushroom caps by their eager push against the medium. I conceived the importance bubbles were to have for us in the dives to come. As long as air boiled on the surface all was well below. If the bubbles disappeared there would be anxiety, emergency measures, despair. They roared out of the regulator and kept me company. I felt less alone.

I swam across the rocks and compared myself favorably with the sars. To swim fishlike, horizontally, was the logical method in a medium eight hundred times denser than air. To halt and hang attached to nothing, no lines or air pipe to the surface, was a dream. At night I had often had visions of flying by extending my arms as wings. Now I flew without wings. (Since that first aqualung flight, I have never had a dream of flying.)

I thought of the helmet diver arriving where I was on his ponderous boots and struggling to walk a few yards, obsessed with his umbilici and his head imprisoned in copper. On skin dives I had seen him leaning dangerously forward to make a step, clamped in heavier pressure at the ankles than the head, a cripple in an alien land. From this day forward we would swim across miles of country no man had known, free and level, with our flesh feeling what the fish scales know.

I experimented with all possible maneuvers of the aqualung — loops, somersaults and barrel rolls. I stood upside down on one finger and burst out laughing, a shrill distorted laugh. Nothing I did altered the automatic rhythm of air. Delivered from gravity and buoyancy I flew around in space.

I could attain almost two knots' speed, without using my arms. I soared vertically and passed my own bubbles. I went down to sixty feet. We had been there many times without breathing aids, but we did not know what happened below that boundary. How far could we go with this strange device?

Fifteen minutes had passed since I left the little cove. The regulator lisped in a steady cadence in the ten-fathom layer and I could spend an hour there on my air supply. I determined to stay as long as I could stand the chill. Here were tantalizing crevices we had been obliged to pass fleetingly before. I swam inch-by-inch into a dark narrow tunnel, scraping my chest on the floor and ringing the air tanks on the ceiling. In such situations a man is of two minds. One urges him on toward mystery and the other reminds him that he is a creature with good sense that can keep him alive, if he will use it. I bounced against the ceiling. I'd used one-third of my air and was getting lighter. My brain complained that this foolishness might sever my air hoses. I turned over and hung on my back.

The roof of the cave was thronged with lobsters. They stood there like great flies on a ceiling. Their heads and antennae were pointed toward the cave entrance. I breathed lesser lungsful to keep my chest from touching them. Above water was occupied, ill-fed France. I thought of the hundreds of calories a diver loses in cold water. I selected a pair of one-pound lobsters and carefully plucked them from the roof, without touching their stinging spines. I carried them toward the surface.

Simone had been floating, watching my bubbles wherever I went. She swam down toward me. I handed her the lobsters and went down again as she surfaced. She came up under a rock which bore a torpid Provençal citizen with a fishing pole. He saw a blonde girl emerge from the combers with lobsters wriggling in her hands. She said, "Could you please watch these for me?" and put them on the rock. The fisherman dropped his pole.

Simone made five more surface dives to take lobsters from me and carry them to the rock. I surfaced in the cove, out of the fisherman's sight. Simone claimed her lobster swarm. She said, "Keep one for yourself, *monsieur*. They are very easy to catch if you do as I did."

Lunching on the treasures of the dive, Tailliez and Dumas questioned me on every detail. We reveled in plans for the aqualung. Tailliez pencilled the tablecloth and announced that each yard of depth we claimed in the sea would open to mankind three hundred thousand cubic kilometers of living space. Tailliez, Dumas and I had come a long way together. We had been eight years in the sea as goggle divers. Our new key to the hidden world promised wonders. We recalled the beginning. . . .

Our first tool was the underwater goggle, a device that was known centuries ago in Polynesia and Japan, was used by sixteenth-century Mediterranean coral divers, and has been rediscovered about every decade in the last fifty years. The naked human eye, which is almost blind under water, can see clearly through watertight spectacles.

One Sunday morning in 1936 at Le Mourillon, near Toulon, I waded into the Mediterranean and looked into it through Fernez goggles. I was a regular Navy gunner, a good swimmer interested only in perfecting my crawl style. The sea was merely a salty obstacle that burned my eyes. I was astounded by what I saw in the shallow shingle at Le Mourillon, rocks covered with green, brown and silver forests of algae and fishes unknown to me, swimming in crystalline water. Standing up to breathe I saw a trolley car, people, electric-light poles. I put my eyes under again and civilization vanished with one last bow. I was in a jungle never seen by those who floated on the opaque roof.

Sometimes we are lucky enough to know that our lives have been changed, to discard the old, embrace the new, and run headlong down an immutable course. It happened to me at Le Mourillon on that summer's day, when my eyes were opened on the sea.

Soon I listened hungrily to gossip about heroes of the Mediterranean, with their Fernez goggles, Le Corlieu foot fins, and barbarous weapons to slay fish beneath the waves. At Sanary the incredible Le Moigne immersed himself in the ocean and killed fish with a slingshot!

There also was a fabulous creature named Frédéric Dumas, son of a physics professor, who speared fish with a curtain rod. These men were crossing the frontier of two hostile worlds.

Two years of goggle dives passed before I met Dumas. He told me how it had begun with him. "One day in the summer of 1938 I am out on the rocks when I see a real manfish, much further on in evolution than me. He never lifts his head to breathe, and after a surface dive water spouts out of a tube he has in his mouth. I am amazed to see rubber fins on his feet. I sit admiring his agility and wait until he gets cold and has to come in. His name is Lieutenant de Vaisseau Philippe Tailliez. His undersea gun works on the same theory as mine. Tailliez's goggles are bigger than mine. He tells me where to get goggles and fins and how to make a breathing pipe from a garden hose. We make a date for a hunting party. This day is a big episode in my undersea life."

The day was important for each of us. It brought Tailliez, Dumas and me into a diving team. I already knew Tailliez.

Undersea hunting raged, with arbalests, spears, spring guns, cartridge-propelled arrows, and the elegant technique of the American writer, Guy Gilpatric, who impaled fish with fencing lunges. The fad resulted in almost emptying the littoral of fish and arousing the commercial fishermen to bitter anger. They claimed we drove away fish, damaged nets, looted their seines, and caused mistrals with our schnorkels.

One day, however, when Dumas was diving he noticed a picturesque individual watching him from a large power boat, a formidable man stripped to the waist. He exhibited a gallery of torsoid tattoos consisting of dancing girls and famous generals such as Maréchal Lyautey and Papa Joffre. Didi winced as

We enter the sea, holding our breaths on goggle dives, in 1936. In 1938 my closest companion, Frédéric Dumas, plunges into the Mediterranean on a bet that he could spear 100 kilos (220 pounds) of fish in a morning.

the individual hailed him, for he recognized Carbonne, the dreaded Marseilles gangster, whose idol was Al Capone.

Carbonne summoned Didi to his ladder and handed him aboard. He asked him what he was doing. "Oh, just diving," Didi said, warily.

"I am always coming out here to take a peaceful rest from the city," said Carbonne. "I like what you are doing. I wish you to conduct all of your activities from my ship."

Didi's patron heard about the fishermen's hatred of divers. It incensed him and, cruising among the fishing boats with his hairy arm flung over Didi's shoulder, he bawled out, "Hey, you fellows—don't forget this is *my friend!*"

We twitted Didi about his gangster, but noted that the fishermen no longer molested him. They diverted their protests to the government, which passed a law severely regulating underwater hunting. Air-breathing apparatus and cartridge-propelled harpoons were forbidden. Divers were required to take out hunting licenses and join a recognized spear-fishing club. But from Menton to Marseilles the shore had emptied of larger fauna. Another remarkable thing was noticed. The big pelagic fish had learned how to stay out of range of weapons. They would insolently keep five feet away from a slingshot, exactly beyond its range. A rubber-propelled harpoon gun, which could shoot eight feet, found the fish a little over eight feet away. They stayed fifteen feet from the biggest harpoon guns. For ages man had been the most harmless animal under water. When he suddenly learned underwater combat, the fish promptly adopted safety tactics.

In the goggle-diving era Dumas made a lighthearted bet at Le Brusq that he could spear two hundred and twenty pounds of fish in two hours. He made five dives within the time limit, to depths of forty-five to sixty feet. On each dive he speared

Above, Dumas breaks a spear on a 50-pound grouper. He goes up for another spear, submerges again, gets his fish, and here ascends with the broken harpoon hanging from the grouper. Below, Dumas wins his bet with 280 pounds of fish in five dives.

and fought a mammoth fish in the short period he could hold his breath. He brought up four groupers and an eighty-pound liche (palomata or leerfish).* Their total weight was two hundred and eighty pounds.

One of our favorite memories is of a fighting liche which probably weighed two hundred pounds. Didi speared him and we went down in relays to fight him. Twice we managed to drag him to the surface in our arms. The big fellow seemed to like air as much as we did. He gained strength as we wore out, and at last the monarch of liches escaped.

We were young and sometimes we went beyond the limits of common sense. Once Tailliez was diving alone in December at Carqueiranne, with his dog Soika guarding his clothes. The water was 52 ° Fahrenheit. Philippe was trying to spear some big sea bass but had to break off the chase when he could no longer stand the cold. He found himself several hundred yards from the deserted shore. The return swim was a harrowing, benumbed struggle. He dragged himself out on a rock and fainted. A bitter wind swept him. He had small chance of surviving such an exposure. The wolfhound, moved by an extraordinary instinct, covered him with its body and breathed hot air on his face. Tailliez awoke with near-paralyzed hands and feet and stumbled to a shelter.

Our first researches in diving physiology were attempts to learn about cold. Water is a better heat conductor than air and has an extreme capacity for draining off calories. Bodily heat lost in sea bathing is enormous, placing a grave strain on the central heating plant of the body. The body must above all keep its central temperature constant. Exposed to cold the body makes a ruthless strategic retreat, first abandoning the skin to cold and then the subcutaneous layer, by means of vasoconstriction of the superficial blood vessels—gooseflesh. If the cold continues to draw off heat, the body will surrender the

* Even where in this book the English equivalent of a fish name is given, it should be remembered that the type may vary from fish of the same family found in our own waters.

hands and feet to conserve the vital center. When the inner temperature drops life is in danger.

We learned that bathers who wrapped in blankets were doing exactly the wrong thing. A covering does not restore heat, it merely requires the central heating plant to burn up more calories to flush warmth into the outer layer. The process is accompanied by severe nervous reactions. By the same token hot drinks and alcohol are useless in restoring surface temperature. We sometimes take a drink of brandy after a hard dive, rather for its depressant effect than with any expectation of gaining warmth from it. We learned that the best way to restore heat is the most obvious one, to get into a very hot bath or stand between two fires on the beach.

We discovered a surprising fact about the practice of coating oneself with grease for cold swims. Grease does not stick to the skin. It washes away, leaving a mere film of oil, which, far from protecting the swimmer, slightly increases the loss of caloric heat. Grease would be acceptable insulation, however, if it could be injected under the skin to simulate the splendid blubber underwear of the whale.

To protect myself from cold I spent days tailoring and vulcanizing rubberized garments. In the first one, I looked something like Don Quixote. I made another which could be slightly inflated to provide more insulation, but there was only one depth in which the suit was equilibrated, and I spent most of the time fighting against being hauled up or down. Another weakness of this dress was that the air would rush to the feet, leaving me in a stationary, head-down position. Finally, in 1946, we evolved the constant-volume dress we use now in cold water. It is inflated by the diver's nasal exhalations, blown out under the edges of an inner mask. Air escape valves at the head, wrist and ankles keep the diver stable in any depth or bodily position. Marcel Ichac, the explorer, found it effective in dives under Greenland ice floes on the recent Paul-Émile Victor Arctic expedition. Dumas has designed a "mid-season" dress, a featherlight foam rubber jerkin which protects for twenty minutes in cold water and leaves the diver all his agility.

Vanity colored our early skin dives. We plumed ourselves at the thought that we late-comers could attain the working depths of pearl and sponge divers who had made their first plunges as infants. In 1939 on Djerba Island, off Tunisia, I witnessed and confirmed with a sounding line a remarkable dive by a sixty-year-old Arab sponge diver. Without breathing apparatus he reached one hundred and thirty feet, in an im-

I feel like Don Quixote in my first homemade protective suit in 1938. It kept me warm in cold-water goggle dives. Right, testing our latest "constant-volume" diving dress under Greenland ice floes is our friend, the Himalayan explorer, Marcel Ichac. (Photo property of **Marcel Ichac.***)*

mersion time of two and one-half minutes.

The ordeals of such dives are only for the exceptional man. As the naked diver sounds through increasing pressure layers, the air in his lungs is physically shrunk. Human lungs are balloons in a flexible cage, which is literally squeezed in under pressure. At a hundred feet down the air in the balloon occupies one-fourth the space it does at the surface. Further down the ribs reach a position of inflexibility and may crack and collapse.

However, the working depth of sponge divers is usually not more than the three-atmosphere strata, sixty-six feet, where their rib cages are reduced to one-third normal size. We learned to go that deep without apparatus. We made sixty-foot dives of two-minute durations, aided by several pounds of belt weights. Under twenty-five feet the weights became heavier in proportion to the compression of the rib cage, so that there was a certain uneasiness about meeting with accidents while weighted to the bottom.

Dumas's skin-diving technique consisted of floating face under water and breathing through a schnorkel tube. When he spotted some attraction below, he would execute a maneuver called the *coup de reins*, literally "stroke of the loins," the technique the whale uses to sound. For a floating man, it consists of bending from the waist and pointing the head and torso down. Then the legs are thrown up in the air with a powerful snap and the diver plummets straight down. Lightning dives require well-trained, wide-open Eustachian tubes to deal with the rapidly mounting pressure.

When we had attained the zone of sponge divers we had no particular sense of satisfaction, because the sea concealed enigmas that we could only glimpse in lightning dives. We wanted breathing equipment, not so much to go deeper, but to stay longer, simply to live a while in the new world. We tried Commandant Le Prieur's independent diving gear, a cylinder of compressed air slung across the chest and releasing a continuous flow into a face mask. The diver manually valved the air to meet pressure and cut down waste. We had our first grand moments of leisure in the sea with Le Prieur's lung. But the continuous discharge of air allowed only short submersions.

The gunsmith of my cruiser, the *Suffren*, built an oxygen rebreathing apparatus I designed. He transformed a gas-mask canister of soda lime, a small oxygen bottle, and a length of motorbike inner tube into a lung that repurified exhalations by

The ancestor of our aqualung diving apparatus was Le Prieur's 1933 compressed-air cylinder. Divers could walk in shallow depths, feeding themselves air from a hand valve. In 1942 the brilliant engineer, Émile Gagnan, and I made the first completely automatic compressed-air apparatus, the aqualung.

filtering out the carbon dioxide in the soda lime. It was self-contained, one could swim with it, and it was silent. Swimming twenty-five feet down with the oxygen apparatus was the most serene thrill I have had in the water. Silent and alone in a trancelike land, one was accepted by the sea. My euphoria was all too short.

Having been told that oxygen was safe down to forty-five feet, I asked two sailors from the *Suffren* to man a dinghy above me, while I dived to the boundary of oxygen. I went down with a ceremonious illusion. I was accepted in the sea jungle and would pay it the compliment of putting aside my anthropoid ways, clamp my legs together and swim down with the spinal undulations of a porpoise. Tailliez had demonstrated that a man could swim on the surface without using arms or legs. I borrowed the characteristics of a fish, notwithstanding certain impediments such as my anatomy and a ten-pound lead pipe twisted around my belt.

I undulated through the amazingly clear water. Ninety feet away I saw an aristocratic group of silver and gold giltheads wearing their scarlet gill patches like British brigadiers. I wiggled toward them and got very close without alarming them. My fish personality was fairly successful, but I remembered that I could swim a great deal faster by crudely kicking my fins. I started chasing a fish and cornered him off in his cave. He bristled his dorsal fins and rolled his eyes uneasily. He made a brave decision and sprang at me, escaping by inches. Below I saw a big blue dentex (bream) with a bitter mouth and hostile eyes. He was hanging about forty-five feet down. I descended and the fish backed away, keeping a good distance.

Then my lips began to tremble uncontrollably. My eyelids fluttered.

My spine was bent backward like a bow.

With a violent gesture I tore off the belt weight and lost consciousness.

The sailors saw my body reach the surface and quickly hauled me into the boat.

I had pains in neck and muscles for weeks. I thought my soda lime must have been impure. I spent the winter on the *Suffren* building an improved oxygen lung, one that would not induce convulsions. In the summer I went back to the same place off Porquerolles and went down forty-five feet with the new lung. I convulsed so suddenly that I do not remember jettisoning my belt weight. I came very near drowning. It was the end of my interest in oxygen.

In the summer of 1939 I made a speech at a dinner party, explaining why war could not come for at least ten years. Four days later I was aboard my cruiser, speeding west under secret orders; the next day at Oran we heard the declaration of war. At our ship line lay a division of Royal Navy torpedo boats, one of which was disabled by a heavy steel cable fouled in her screw. There were no navy divers at Oran. I volunteered to make a skin dive to survey the situation.

Even the sight of the screw did not cool my ardor: the thick wire was wound six times around the shaft and several times around the blades. I called on five good skin divers from my ship, and we dived repeatedly to hack away the cable. After hours of work clearing the propeller, we crawled back on our ship, barely able to stand. The torpedo boat sailed out with its division, and as it passed, the crew turned out in a line at the rail and gave three cheers for the crazy Frenchmen. That day I learned that heavy exertion under water was madness. It was absolutely necessary to have breathing apparatus to do such jobs.

Later in the war while I was working for Naval Intelligence in Marseilles against the occupying powers, my commander insisted that I continue diving experiments when my duties permitted. Diving helped camouflage the secret work. I tested the Fernez diving apparatus, which consisted of an air pipe from a surface pump. The pipe was carried across the diver's face to a duck-beak valve which released a constant flow of pumped air. The diver tapped the flow with a mouthpiece, sucking the air he needed. It was the simplest diving gear ever designed. It tethered a man to the surface and unnecessarily wasted half the air, but at least it did not use treacherous oxygen.

I was enjoying the full breaths of the Fernez pump one day at forty feet when I felt a strange shock in my lungs. The rumble of exhaust bubbles stopped. Instantly I closed my glottis, sealing the remaining air in my lungs. I hauled on the air pipe and it came down without resistance. The pipe had broken near the surface. I swam to the boat. Later I realized the danger I had faced. If I hadn't instinctively shut the air valve in my throat the broken pipe would have fed me thin surface air and the water would have collapsed my lungs in the frightful "squeeze."

In testing devices in which one's life is at stake, such accidents induce zeal for improvement. We were working on defenses against broken pipes one day with Dumas seventy-five feet down, breathing from the Fernez pipe. I was in the tender, watching the pipe, when I saw it rupture. Dumas was trapped in pressure three times greater than the surface. I grabbed the pipe before it sank and reeled it in frantically, ill with suspense. I could feel heavy tugs from below. Then Dumas appeared, red-faced and choking, his eyes bulging. But he was alive. He, too had locked his glottis in time and had then climbed the pipe hand over hand. We worked on the gear until it operated more reliably, but the pump could take us no further. It fastened us on a leash and we wanted freedom.

We were dreaming about a self-contained compressed-air lung. Instead of Le Prieur's hand valve, I wanted an automatic device that would release air to the diver without his thinking about it, something like the demand system used in the oxygen masks of high-altitude fliers. I went to Paris to find an engineer who would know what I was talking about. I had the luck to

Overleaf, Philippe Tailliez, of the Undersea Research Group, swims in complete freedom with an aqualung, receiving breaths automatically from the demand regulator.

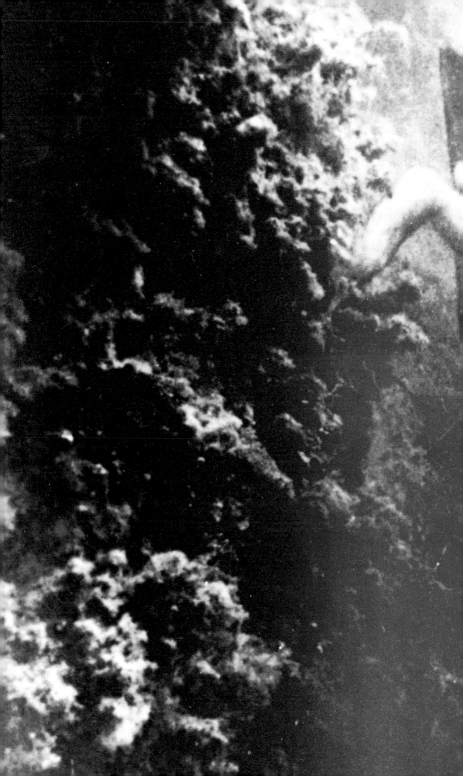

meet Émile Gagnan, an expert on industrial-gas equipment for a huge international corporation. It was December, 1942, when I outlined my demands to Émile. He nodded encouragingly and interrupted. "Something like this?" he asked and handed me a small bakelite mechanism. "It is a demand valve I have been working on to feed cooking gas automatically into the motors of automobiles." At the time there was no petrol for automobiles and all sorts of projects were under way for utilizing the fumes of burning charcoal and natural gas. "The problem is somewhat the same as yours," said Émile.

In a few weeks we finished our first automatic regulator. Émile and I selected a lonely stretch of the river Marne for a test dive. He stood on the bank while I waded in over my head. The regulator furnished plenty of air without effort on my part. But the air rushed wastefully out of the exhaust pipe in the fashion of the Fernez gear. I tried standing on my head. The air supply almost ceased. I couldn't breathe. I tried swimming horizontally, and the air flowed in a perfectly controlled rhythm. But how were we going to dive if we couldn't operate vertically?

Chilled and disappointed, we drove home, analyzing the regulator's reason for such tricks. Here it was, a miracle of design, the first stage efficiently reducing one hundred and fifty atmospheres to six atmospheres, and the second control stage rationing that to breathing density and volume. Before we reached Paris we had the answer.

When I was standing up in the water the level of the exhaust was higher than the intake and that six-inch difference in pressure allowed the air to overflow. When I stood on my head, the exhaust was six inches lower, suppressing the air flow. When I swam horizontally, the exhaust and intake were in the same pressure level and the regulator worked perfectly. We arrived at the simple solution of placing the exhaust as close as possible to the intake so that pressure variations could not disrupt the flow. The improvement worked perfectly in a tank test in Paris.

Chapter Two **Rapture of the Deep**

T HE first summer in the sea with the aqualung was a memorable time. It was 1943, the middle of a war in my occupied country, but in the delight of diving we thought nothing of these improbable circumstances. Living in Villa Barry were Dumas; Tailliez, his wife and child; Claude Houlbreque, the cinematographer, and his wife; Simone and I and our two children. Frequent guests were Roger Gary and his wife. He was an old personal friend, the director of a Marseilles paint factory. To the occupying troops we must have seemed a wistful holiday party.

The first requirement of diving was to feed our band of twelve. Tailliez went to the country and returned with five hundred pounds of dried beans, which we stored in the coal bin and ate for breakfast, lunch and dinner, with an occasional maggot to break the monotony. Diving burns more calories than working in a steel mill. We managed to get "heavy worker" ration cards which allowed a few grams of butter and more bread. Meat was a rarity. We ate few fish. We calculated that in our weakened physical conditions an undersea hunter would burn more calories chasing and fighting a fish than his portion of the game would restore.

That summer we logged five hundred dives with the aqualung. The more accustomed we became to it, the more we feared the sudden catastrophe which the oxygen lung and Fernez pump accidents had led us to expect. The thing was too easy. Every instinct insisted that we could not so flippantly invade the sea. An unforeseen trap awaited in the deep, any day now, for Dumas, Tailliez or me.

Our friends ashore listened to our undersea reports with

maddening boredom. We were driven to making photographs to reveal what we had seen. Since we were always on the move downstairs we began with motion pictures. Our first camera was an obsolete Kinamo I bought for $25. Papa Heinic, a Hungarian refugee, ground a fine lens for it; Léon Veche, machinist of the torpedo boat *Le Mars*, built a watertight case. We could obtain no 35 mm. movie film in wartime. We bought up fifty-foot rolls of Leica film and spliced the negatives together in a darkroom.

One of the film locations was Planier Island, in the main roadstead of Marseilles, the site of a famous lighthouse which the departing Germans wantonly destroyed in 1944. Under Planier on a treacherous rocky plateau lay a five-thousand-ton British steamer, *Dalton*, fifty feet under water at the bow and slanting steeply down to the stern. The ship had met an interesting fate.

Under charter to a Greek navigation company, the *Dalton* left Marseilles on Christmas night, 1928, with a cargo of lead. Attracted to Planier Light like a mosquito to a lamp, the vessel steamed straight into the island, struck heavily and foundered. The light keepers climbed down the rocks and rescued all hands. According to the keepers' story, every one of them was drunk, from cabin boy to skipper. Holiday spirit had overcome them indiscriminately.

With a permit from the lighthouse administration, Tailliez, Gary, Dumas, Houlbreque and I landed from the weekly supply boat with aqualungs, spears, crossbows, cameras, air compressor and food. The lighthouse people were on edge, wondering when the Nazis were going to destroy the light, or when a British submarine would surface at night and seize the island.

We walked down the stone stairs into the sea and swam to the *Dalton*'s forepeak. The depth was dramatically asserted by an abrupt rock wall and by a "wedging sense" in the ears as one journeyed down. Sounding headlong into pressure sometimes seems as though one is driving in one's head like a wedge,

The wheel of the Greek steamer, Dalton, *90 feet down near Planier Island. The* Dalton *went down on Christmas Eve forty-five years ago.*

then the ears loosen with swallowing and one feels at ease.

We passed the butting prow and followed the long flanks down into the blue, past half-collapsed sides to a buckled deck with a yawning cargo hatch. We tilted down through the hatch, blinking and expanding our pupils in the dark. The hold was a great tunnel leading down. It was paved with sand and iron plate and, where the *Dalton* had broken apart in the crash, a great jagged tunnel mouth opened on still deeper sea. I hung in the dark tunnel and watched my mates coming down past jagged barbs of iron. They issued locomotivelike puffs of bubbles.

Amidships torn railings made a jungle in which dentex and black bream circled like birds. Beneath the shattered bridge we came to the main engine-room control wheel, deeply encrusted and almost obscured by tiny pomfrets, "the flies of the sea." The bulkheads were geometrically patterned in foliage where pipes and gauges were buried. We were a hundred feet down, visiting the zone of the unforeseeable. We hung there, looking down the slope and saw, framed in the hull mouth, down another dune, the severed stern quarters of the *Dalton*. The structure lay thirty feet deeper, undamaged and beckoning, with two masts still standing on it.

We had started aqualung diving with no plans for deep descents. We had wanted to spend some time at sixty feet, but the sea lured us down. Now we were in the risky seventeen-fathom layer. Where did the depth limit lie? Perhaps it was in the tantalizing open dune between the two halves of the *Dalton*. We decided we had better surface and think it over.

On the island we faced a supreme commonplace problem, how to nourish ourselves. A diver needed four pounds of meat

When we built the first aqualungs, Dumas stayed down and speared fish by the hundreds. Here he is impaling a grouper in 1943 in the wreck of the Dalton, *90 feet down. Below, Dumas on the way to the surface with a 60-pound grouper on his spear, the fish he took in the wreck of the* Dalton.

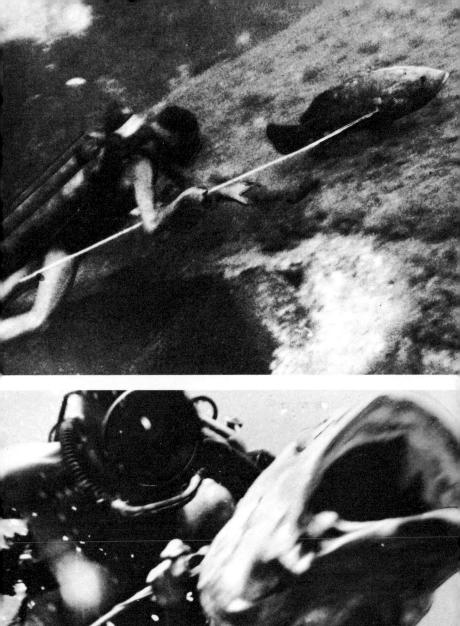

a day. Tailliez and Dumas decided to defy the law that fish could not replace the calories expended in spearing them. The big groupers around the *Dalton*'s forepeak had never been hunted and all but stood still for the spears and bolts of Dumas. We made caldrons of *bouillabaisse.* We split the fish but did not gut them for the pot. The head, eyes, brains and entrails gave indescribable nuances of flavor, lost in sophisticated cooking. One did not have to eat an eye, of course, but the broth had the superb juices of the offal that is thrown away by all but primitive peoples.

The grouper we hunted was a large species called the merou, virtually unknown in the Provençal markets until goggle divers went down and speared them. Commercial fishermen had seen them through glass-bottomed buckets, but could not net them. They had been rarely hand-lined. When hooked, the merou goes into its rock crevice and puts up a stubborn siege defense by erecting its spines and bracing itself in the hole. Arabs trick it out by dangling an octopus outside and giving a well-timed jerk, which sometimes brings up a merou, but more often does not. A clever means of dislodging the hooked grouper is to let a heavy weight slide down the line. When the messenger bangs the merou's muzzle, it relaxes the spines at the moment of shock. Hauling at that instant may free the fish or at least move it a few inches forward. Sending down messengers one by one, and patiently timing the jerks, may overcome the defense.

One of Didi's merou suppers, a forty-pound individual, gave him a hair-raising chase. He encountered it near the *Dalton.* The merou demonstrated the remarkable speed with which fish learn to deal with undersea hunters. It maintained a safe distance from the harpoon gun, and at last broke off and made for its castle. Dumas saw his last chance and fired. The harpoon went through the merou. The big fish took off, towing Didi. It

Tailliez tries to knife a grouper, which is bristling its dorsal spines in wariness.

dived under the hull, pulling Dumas into an unhealthy situation where his chest scraped the sand and his aqualung cylinders rang on the ponderous hull above. The siege situation was reversed. Now the man was being wedged into a crevice by the fish. The merou passed out of sight, hauling Didi further into the trap. In the almost total darkness Didi could see only the cork float on his harpoon line. The cork became lodged between rock and iron, mooring the fish.

Dumas cut away the line and crawled out backward, praying that the corroded hull would stand the banging of his air tanks. Above him the plates had been eaten through in many places. He wriggled clear and sized up the situation. He decided to get the bold fish if he could. He swam topside and down into the hull and located his float in a jagged pit. His first pull re-awakened the pain of the fish. With a powerful tug, it pulled him down into the maze again. He went hand over hand down his line and got a hold of the harpoon.

There was a furious fight in the dark, in sand clouds stirred up by the struggling bodies. At last Didi got control and turned the merou toward the exit. Then he held on to the spear like a tiller and the fish gave him a fast ride through the winding maze to the open floor. It was a hard way to buy fish, but we were hungry.

We nerved ourselves to the inescapable—we had to go on down to the stern section of the *Dalton*. There was no other way to find the limitations of the lung. We planed down the long belly and broke into the clear ominous gap below which the afterwork lay in one hundred and thirty feet of water. In the crystalline sea the site had a magical appearance. Objects no longer had shadows. Masts, iron plates and men hung big and soft in light that poured from everywhere.

The afterdeck planking was gone, exposing a tangle of steel beams and ribs. There was no familiar soft green and brown algae. The biological husk was hard and sharp. On the quarter-deck there was a strange structure that looked like the Ark of the Covenant that is carried in the streets on Saints' Days. It was the old-fashioned cockpit, above which rose the emergency wheel with missing spokes, standing against an insect horde of little black fish.

We swam hesitantly to the taffrail and looked down at the dull sand falling away to an amorphous horizon. We seemed to be as comfortable as we had been at fifty feet. We were acquiring an extra sense of the sea, a sort of autodiagnosis of depth. We meditated on our feelings, trying not to imagine symptoms that were not there.

Before we dropped off the taffrail we instinctively palpated the water to be sure it was there to support us as we abandoned ship. Over the rail we went and arrived at the bottom of the sea. The *Dalton*'s screw blades were half buried in agitated sand that seemed to have been carved by their last convulsions. Beneath the propeller arch, deeper than we had ever been, we felt no unusual discomfort, but an extra exertion made us gasp. If we swam too fast or tried to handle heavy objects the respiratory cycle was broken.

We kicked off for the surface, laying triple plumes of smoke above the hulk, and passed into the rock slope beneath the stone stair of Planier Light. Suddenly my mask went out of focus and scintillating scotomata flashed across my eyeballs. I held on to a rock and closed my eyes. This was the punishment of the sea. At last I took a look at what life was like. It was cheerfully normal. Wave prisms played lazily on the rocks. My companions were gone. I swam up and sat down on the stone steps, and the Mediterranean laughed in the sun. Later I learned that the phenomenon occurs during decompression when the ear, which contains the organs of equilibrium, becomes congested and gives the diver a moment of vertigo and falling stars. The effect is of no consequence.

Dumas believed that the lung could take us deeper, after we had safely returned to twenty-two fathoms several times the first summer. He decided to try the limit a man could reach in a carefully controlled experimental dive. We figured that he would not be down long enough to incur an attack of "the bends."

We already knew something about the bends, from the

pioneering work of the French scientist, Paul Bert, in the eighteen-seventies and from the advanced studies of later British and American physiologists. The bends, or caisson disease, is a painful, crippling and sometimes fatal affliction of divers, of which the first notable medical observation was made on the sandhogs who worked in dry pressurized shafts to dig the pier excavations of the Brooklyn Bridge. The workers often came up in tortured bodily positions which reminded their mates of a feminine posture fad of the moment called "the Grecian bend." Ever since then this terrible and easily averted accident has been called "the bends."

It is caused by the fact that a diver in pressure is breathing multiples of nitrogen, an inert gas which constitutes 78 per cent of the atmosphere and does not entirely pass away in a diver's exhalations. Instead it goes into solution in the blood and gristle. When the diver rises into lesser pressure, the nitrogen comes out of compression and becomes froth, on the same principle as opening a bottle of champagne. The CO_2 in champagne, which has been under pressure by the cork, expands theatrically when liberated. So does the nitrogen of the diver's body when he passes into lighter water pressure. In mild cases the froth gives him pains in the joints. In severe cases the nitrogen bubbles can clog the veins, cut off the spinal ganglia or cause instant death by heart embolism.

On an October afternoon in 1943 we arrived in a Mediterranean fishing village to rendezvous with persons involved in the test. A hundred-meter length of knotted rope lying along the jetty was under examination by Monsieur Mathieu, the harbor engineer, and Maître Gaudry, *huissier*. This French functionary is licensed by the Republic as a bailiff, unimpeachable witness and investigator. His testimony is accepted without challenge in any court of law. The engineer and the *huissier* methodically counted and measured the metric knots in the rope along which Frédéric Dumas was to descend into the sea.

Two launches, full of witnesses, accompanied the condemned man to sea. The first launch towed the second, in which were

Didi and I, embarrassed by the attentions of the crowd. We had talked over all conceivable problems of the dive, and Didi had himself weighed, catalogued everything that could happen and was ready for it. The plunge was well planned. He would submerge in the clear calm water, wearing a factory-new aqualung and a heavily weighted belt, and descend feet first without undue exertion along the knotted rope to the greatest depth he could reach. There he would remove his weights, tie them on the rope, and speed to the surface. When the line was brought up it would show what depth he had attained. After the terrors of thinking about it, Didi considered the plunge a formality.

The towboat anchored in two hundred and forty feet of water. The sky was clouded and an early autumn wind drove a muddy white-crowned chop past our gunwales. The air was raw. As Dumas's safety man, I entered the water first and was swept away from the launch. I swam hard to get back to the ladder and had to struggle to stay alongside. Didi came into the water. The launch skipper was distressed at the sight of men abandoning a vessel in such a sea and ran around, hurling lines to us. Dumas saluted his gallantry and sank. He did so unwillingly, as he was overweighted. Underwater he discovered that when he turned his head to the left it pinched off his air intake hose. I swam to catch the knotted rope as it was thrown overboard. I clutched the rope, out of breath, with the big dive not yet started. Dumas went under again.

I looked down and saw Didi sinking under his weights and swimming with both arms and legs against the sweep of current to gain the shotline. When he caught it a flume of air came out of his regulator, a sign of exhaustion. He rested on the rope for a moment and then lowered himself rapidly hand under hand into the turbid, racing sea.

Still panting from the fight on the surface, I followed him toward my sentry post a hundred feet down. My brain was reeling. Didi did not look up. I saw his fists and head melting into the dun water.

Here is how he described the record dive:

"The light does not change color as it usually does underneath a turbid surface. I cannot see clearly. Either the sun is going down quickly or my eyes are weak. I reach the hundred-foot knot. My body doesn't feel weak but I keep panting. The damned rope doesn't hang straight. It slants off into the yellow soup. It slants more and more. I am anxious about that line, but I really feel wonderful. I have a queer feeling of beatitude. I am drunk and carefree. My ears buzz and my mouth tastes bitter. The current staggers me as though I had had too many drinks.

"I have forgotten Jacques and the people in the boats. My eyes are tired. I lower on down, trying to think about the bottom, but I can't. I am going to sleep, but I can't fall asleep in such dizziness. There is a little light around me. I reach for the next knot and miss it. I reach again and tie my belt on it.

"Coming up is merry as a bubble. Liberated from weights I pull on the rope and bound. The drunken sensation vanishes. I am sober and infuriated to have missed my goal. I pass Jacques and hurry on up. I am told I was down seven minutes."

Didi's belt was tied off two hundred and ten feet down. The *huissier* attested it. No independent diver had been deeper. Yet Dumas's subjective impression was that he had been slightly under one hundred feet.

Didi's drunkenness was nitrogen narcosis, a factor of diving physiology which had been studied by Captain A. R. Behnke, U.S.N., several years before. In occupied France we knew nothing of his work. We called the seizure *l'ivresse des grandes profondeurs* (rapture, or "intoxication," of the great depths).

The first stage is a mild anesthesia, after which the diver becomes a god. If a passing fish seems to require air, the crazed diver may tear out his air pipe or mouth grip as a sublime gift. The process is complex and still an issue among diving physiologists. It may derive from nitrogen oversaturation, according to Captain Behnke. It has no relation to the bends. It is a gaseous attack on the central nervous system. Recent laboratory studies attribute "rapture of the great depths" to residual carbon dioxide retained in the viscosity of nerve tissues. U.S. Navy test dives have shown that the strange joy does not occur to deep divers in whose air supply nitrogen has been supplanted by helium. The world's only industrial helium well is in the United States, protected by a rigid law, so that foreign experimenters may not utilize it. Hydrogen, another gas lighter than air, may be as effective as helium, but it is explosive and hard to handle. The Swede, Zetterstrom, used hydrogen in his air supply on a spectacular deep dive, but he died during decompression, due to personnel failure on the surface, before he could contribute much data to the question.

I am personally quite receptive to nitrogen rapture. I like it and fear it like doom. It destroys the instinct of life. Tough individuals are not overcome as soon as neurasthenic persons like me, but they have difficulty extricating themselves. Intellectuals get drunk early and suffer acute attacks on all the senses, which demand hard fighting to overcome. When they have beaten the foe, they recover quickly. The agreeable glow of depth rapture resembles the giggle-party jags of the nineteen-twenties when flappers and sheiks convened to sniff nitrogen protoxide.

L'ivresse des grandes profondeurs has one salient advantage over alcohol—no hangover. If one is able to escape from its zone, the brain clears instantly and there are no horrors in the morning. I cannot read accounts of a record dive without wanting to ask the champion how drunk he was.

The funniest story on pressure I have heard was told by Sir Robert H. Davis, the diving historian and inventor of the first submarine escape apparatus. Years ago during the construction of a tunnel under a river, a party of politicians went down to celebrate the meeting of the two shafts. They drank champagne, disappointed that the wine was flat and lifeless. It was under depth pressure, of course, and the carbon dioxide bubbles remained in solution. When the town fathers arrived at the surface the wine popped in their stomachs, distended their vests, and all but frothed from their ears. One dignitary had to be rushed back into the depths to undergo champagne recompression.

Today, a decade after our hesitant penetration of the one hundred and thirty foot zone, women and old men reach that depth on their third or fourth dive. There is a familiar sight on the Riviera in the summer, an aqualung instruction truck, operated by a Monsieur Dubois, who rents lungs and furnishes instruction to anyone who would like to visit the bottom. Hundreds of people put on the lung and plunge confidently. Recalling the ominous struggles Philippe, Didi and I had, my pride in M. Dubois's outfit is not unmixed with resentment.

Chapter Three **Sunken Ships**

T O GO back a little, one night in November, 1942, in our apartment in Marseilles, Simone and I were awakened by planes flying eastward. I tuned in Radio Geneva. Hitler had broken his pledge and invaded the Toulon naval base. The French fleet was destroying itself in a roar of explosions and flame. The announcer's voice broke as he read the roll of ships, which included the *Suffren* and the *Dupleix*, on which I had served. Simone and I wept by the radio, far from the people and ships we loved, feeling a bitter exile.

After the Germans came the Italians, who took over the dockyard, scavenging and destroying. I cannot forget the Italian torches emasculating the gun barrels of the battleships. Sunken ships preyed on our minds. As we planned the diving program for next spring, Dumas talked about nothing but wrecks. We decided to make a film of sunken ships. Mussolini's employees still occupied the South of France. The Italians were in control and would not give us permits to accompany fishing boats. In vain I showed my *ordre de mission* from the International Committee for the Exploration of the Mediterranean, whose former chairman had been the Italian admiral, Taon di Ravel. If we swam out further than the bathers, sentries shot at us, either in fun or anger, I never knew which.

It was different when the Germans took over. When I showed my *ordre de mission*, even the most brutal-looking Hitlerite was impressed. The word *kultur* had a magic effect on them and we could work without much bother. They never inquired what we were doing, which was lucky for us. We learned later that the German Admiralty had been spending millions of marks to develop military frogman equipment and some of

IDENTITÉ DU TITULAIRE

Nom et Prénoms _COUSTEAU, Jacques Yves_

Né le _11 Juin 1910_ à _Saint André (Gironde)_.

Adresse : _48 Avenue de la Motte - Picquet - Paris (15ͤ)_.

Nationalité : _Française_

exerçant { dans l'Établissement } ~~moins de 30 heures~~
{ ~~à son compte~~ } plus de 30 heures
(rayer, les deux mentions, inutiles)

le métier de _Producteur - réalisateur de films culturels_

depuis le _24 Décembre 1942_

Inscrit dans la catégorie _Direction_.

sous le № _1_

Le Titulaire

CERTIFICAT DE TRAVAIL Mod. 1

délivré sous la responsabilité de :

Monsieur _Cousteau, Jacques Yves_

Qualité : _Directeur des_

(l'Employeur indiquera son titre, la raison sociale et l'adresse de son Établissement)

Films Scientifiques J. Y. Cousteau
48, Av. de la Motte - Picquet, Paris (15ͤ)

A _Paris_, le _11 Juin_ 1943

Signature et cachet du Responsable :

J. 37280-43. (B)

T. S. V. P.

The official wartime work card that saved me several times from deportation to Germany. The first side was made out by my employer, Jacques-Yves Cousteau, producer of scientific films. The other side said that he employed Jacques-Yves Cousteau to direct the films.

36

its experimental teams may have been diving near us. We were diving in a thirty-fathom range, while the various naval frogmen with oxygen rebreathing equipment were confined to seven fathoms. But, of course, oxygen lungs had a considerable military advantage over the aqualung in that they sent up no telltale bubbles.

Locating sunken ships was harder than daydreaming about them. The majority of wrecks, which lay in dark filthy harbors or in straits with turbid water and strong currents, were of no interest. Ships that lay in clear water were suitable, but very hard to find. No charts or publications gave accurate positions and, in fact, such information was rarely known by the most interested parties, the shipowner, insurance company, or government bureau. The only way to obtain clues was to sift carefully the stories of salvage contractors, fishermen and helmet divers.

We entered the chase with the help of Auguste Marcellin, a leading salvage contractor in Marseilles. He gave us several wreck locations and lent his tenders and crews for reconnaissance dives. We questioned fishermen in waterfront bistros for tales of wrecks. They had one method of detecting them. If their nets had fouled on something, it was a wreck, *mais certainement*. We dived in many of these places and often found the nets festooned on rocks.

Jean Katsouyanis from Cassis and Michel Mavropointis of Toulon were two retired helmet divers with whom we yarned away many a fascinating evening. They had spent their lives groping around for ships, sponges, red coral and violets. The violet is a strange delicacy cherished only in Marseilles. It is rocklike and lies on stony seabeds, pumping water in and out for nourishment. When disturbed it contracts and fastens tight to a stone, so that a diver must grab very quickly to catch one. Elderly gourmets lurk in the harbor streets, looking for the rare and expensive violets which are hawked by noisy *marchandes*. The connoisseurs cut them open in the street. Inside is a pulpy yolk of most unappetizing appearance—vivid yellow with flecks of red and violet. Holding up the shell the fancier shovels the core in his mouth with his thumb. I tried one. It was like eating iodine. The violet is reputed to cure tuberculosis and increase sexual vigor. Dumas ate fifteen one time to test their powers and reported next morning that he had noticed no effect.

These old Greek divers had worked all over the Mediterranean, off Libya, Greece, Tunisia, Algiers, Spain, Italy and France. They related thrilling tales of fights with big moray eels, of getting lost in black forests beneath the sea. Through their helmet windows they had seen naked divers picking sponges. They explained the physiology of naked diving, and we didn't know whether to laugh or explode. "The man's skin is covered by thousands of tiny bubbles," went the thesis. "They protect him against pressure. If he bumps into something, the bubbles fall off and the man drops dead."

The old fellows had half-crippled arms and legs from "pressure strokes"—actually due to the bends. They considered themselves lucky to be alive. When they were young, half the divers on the big Tunisian sponge harvests were disabled or killed each year by "pressure strokes."

We met a party of Greek professionals at sea off Corsica. In old patched suits and dented helmets they leaped into the water and dropped a hundred and seventy feet in seconds. After ten or fifteen minutes, they came up slowly, but showed complete ignorance of stage decompression, which for a dive of that depth and duration requires an ascending diver to halt ten feet below the surface for nine minutes in order to pass off accumulated nitrogen. When they got out of their impressive suits, they were twisted little men, warped by the bends. They were paid off handsomely for the jewelers' red coral they were gathering. They limped to the bistros and blew the money on drinks and dice.

They assured us that although they were semi-cripples on land, when they returned to the world of pressure, they recovered their agility as in a fountain of youth. The first "pressure stroke" cut them off from land and condemned them to the sea, and each new seizure bound them closer to her. The relief they felt was, of course, the support of water, which eased their palsy.

The sea changes the Greek divers and it profoundly changes the sunken ship. Rust advances under the paint. Weeds and animals come and live on it. From afar it seems to be a rock. Then comes a tremor of recognition. It is a ship that has lost its pride.

The first wreck we visited, before we dived to the *Dalton*, was one of the suicides of Toulon, an ocean-going tug, which

Old and new ways of salvage. Dumas, second from right, joins a veteran conventional diver in cutting up the sunken passenger steamer, Chellah, *near Marseilles.*

lay in forty-five feet of clear water in the outer fairway. It was selected by a Genoese salvage diver named Gianino as a survey job for the Italian Navy. We accompanied him in the role of amateur enthusiasts who wished the privilege of filming him at work.

In eight months luxuriant weeds had thickened the rigging and spars and upholstered the hulk like a floral float in the Nice carnival. Black mussels grew on the ventilators and rails like funeral ornaments. There were many fish about, mainly sea bass, which did not seem to regard us as anything unusual.

Gianino was elated at the opportunity to show his skill. But a helmet diver can hardly walk. He gathered himself for clumsy leaps, raising whirlwinds of dust and algae. The water swirled around him, broadcasting eggs and shoots, enveloping him in his own dust. For us whose investigations depended on never disturbing the bottom, Gianino's leaden feet were disastrous. Inspired by the camera he enacted fantasies. He leaned down dramatically and clutched a red starfish to his breast. We "filmed" long sequences of his underwater *guignol*. We didn't tell Gianino that we had forgotten the camera at Toulon and had weighted the housing with a heavy wrench, in order not to disappoint him.

He lifted the engine-room hatch, tied it back with a rotten cord, valved half the air in his suit and dropped into the hold. Didi's pride was challenged. He had never seen a sunken ship, much less penetrated one, but he followed Gianino down. Didi thought about the rotten cord and returned. Gianino reinflated his dress and soared out like a barrage balloon. He was master of the vertical, but was almost unable to maneuver horizontally. Didi swam lazily along the deck to the forepeak and came upon the standing forecastle hatch. I watched him cautiously open the door, hesitate a moment and plunge in, like a man diving into an ink bottle. In a moment his flippers reappeared and he backed out. Later Didi would have smiled at such timidity.

To one who glided easily across the moss-covered deck,

nothing was wood, bronze or iron. The ship's fittings lost their meaning. Here was a strange tubular hedge, as though trimmed by a fancy gardener. Didi reached into the hedge and turned a wheel. The cylinder rose smoothly. It was a gun barrel. Steel mechanisms last a long time in the sea. We have seen Diesel motors and electric generators brought up in excellent condition, after three years. They must be immediately taken apart and rinsed in soft water, for heavy corrosion starts as soon as the machine hits the air.

Our first old wreck was the battleship *Iéna*, sunk in gun experiments before the first Great War. It was so terribly riddled and wasted in three decades that it seemed some tormented shape the sea had invented. It was impossible to visualize a ship. The remaining plates collapsed when we touched them. In a few more years there will be nothing of the *Iéna*. Iron ships crumble away in the lifetime of a man.

Some years before the recent war the four-thousand-ton freighter *Tozeur* was anchored at Estaque near Marseilles when a treacherous mistral rose, tore out her anchors and flung her against Frioul rock in the outer roads. The prow emerged from the surf and the masts poked up with a slight starboard list. The stern lay sixty-five feet down. The *Tozeur* became our academy of the art of exploring wrecks. She made a path from air to water and practically led us down by the hand. Her intense close-cropped blanket of marine organisms did not remove her ship's look. She was an idealization of a wreck, one of the few that really looked like the sunken ship of a schoolboy's dream.

The *Tozeur* was as treacherous as she was receptive, and taught us much about the actual perils of wrecks. Many of her surfaces were adorned with a nasty little animal, "dog's teeth," a razor-edged incisor clam, which may be venomous as well as sharp. When a swell brushed our near-naked bodies against the ship, we were dealt some scattered cuts. Usually underwater gashes are painless. The sea knows no difference between blood and water. The composition of the two is remarkably alike. But the dog's teeth stung. We also had sudden nose-to-nose encounters with camouflaged scorpion fish, ugly as toads. To us they were no danger, although classified as venomous.

The wood of the wreck was almost crumbled away, but the ironwork was hardly rusted. The bronze hardware was termite-riddled by galvanic effects. Although the water round about was clear, the holds were full of dirty yellowish water. Once a raging mistral cooled the water so that we had to lay off diving for three days until the sea warmed up. We went into the balmy water and passed down into the holds. We got out immediately. The icy mistral water was piled in the holds like a refrigerant.

The *Tozeur* was a fine movie studio. We made long sequences of the handsome ruin for our movie, *Épaves* (*Sunken Ships*). We explored the ship thoroughly and gained technique and confidence. Didi penetrated the engine-room hatch and Philippe and I followed him down. The bulkhead openings were arched in the fashion of cloisters, and there were other effects that induced a religious feeling—the weeds that grew like lichens in a damp chapel and light filtering down as though from clerestory windows. We followed Didi through the iron abbey to the intricate stairwell where decks broke the flights of ladders leading down. Each deck removed more light, shut another door between us and the sun and air. We paused carefully on a stair landing and peered along a shadowy gangway at remote blue lights opening on the sea. We did not feel ready to swim through the tunnels to those lights.

We went down the first flight. There was now iron plate between us and the surface. Didi plunged down another flight and we trailed him. We swam with care, not touching anything. This might be another crumbling card house like *Iéna*. Suddenly the ship reverberated with a tremendous bang. We

Above, Dumas sails down across a barnacle-grown winch of the sunken Tozeur. *Below, Dumas, belowdecks in the cloistered ruin of the* Tozeur, *swims sideways through an engine-room door fifty feet down.*

stiffened and looked at each other. We waited and nothing happened. Didi went down another flight. There came a second loud bang, and a series of them. We convened around Tailliez who grunted, "The swell." That was it. The shallow wreck was working in the swell, popping a rivet or creaking a plate. We swam into the engine room in almost total darkness, and felt that we had accomplished enough for the day.

In the aftcastle we had found a big brilliant bubble, a carboy of some fluid that had not been invaded by the sea. Didi brought it up and gave it to Simone. She smudged some of the liquid on her palm and sniffed it. "It's very good prewar Eau de Cologne," she said.

Didi hunted undersea treasures in the wreck. He snatched electric-light bulbs that still worked, and a pair of mismatched sea boots, while scorpion fish watched without doing their proper guard duty. Under the bridge we found the captain's bathroom. Didi swam down and lay in the bathtub. It was

Dumas peers at me through a porthole of the sunken steamer Tozeur.

quite lifelike — a near-naked man in a bathtub. I almost lost my mouthpiece laughing.

Auguste Marcellin lent us his salvage tender and a crew of hard-bitten helmet divers to film a sequence of oxyacetylene torches cutting up a wreck. The helmet boys decided to give the sports a hazing. They sailed out in a mistral and deliberately moored along the waves so that the small *tartana* rolled violently. A helmeted man leaped into the water and came up with a basket of big bitter mussels from the wreck. He said with a sly grin, "You fellows look thin. I'm going to give you a real meal before we go down." Eating is not the smartest thing one can do before a dive, but we opened the mussels with our daggers and ate the iodine-flavored dainties with apparent relish. The helmet men sat by, watching us. We finished the mussels. Our friend then said, "Here, have some wine and garlic bread." We ate the bread and drank the wine. They laughed, cracked jokes and accepted us. The gustatory feat was much more impressive than our ability to swim under the sea.

A helmet diver started a cutting torch, gripped the burning jet between his feet and jumped into the water. We swam down with him, watching the bubbles streaming off the red flare. He applied the torch to a steel girder and the flame grew, throwing off a rain of melted steel pellets. The disturbed water throbbed against our chests.

While we were working on the *Dalton* Didi gathered a lot of curious loot. On the soil inside the hull he found stacks of crockery, silverware, glass bejeweled with corals, and a large crystal bowl. In this rent and twisted ruin of iron the dishes lay as neat and unbroken as though they were displayed on the gift table at a bourgeois wedding. One day Didi found a midden of *ouzo* bottles and thin Metaxas brandy bottles, all empty. They had been consumed the night the *Dalton*'s carefree crew got drunk and lost their ship on the rocks. It was the scene of the eternal morning-after.

The ship's compass was frosted with coral. We peered in and saw the only animate thing left of the *Dalton* — the compass rose in its alcohol bath, still obeying the distant attraction of the pole. Tailliez took several ship's lanterns as souvenirs, but Didi was insatiable. He sawed off the oaken bridge wheel and dived repeatedly for dishes and silver. We suspected that he was collecting household gear for a wedding he had failed to mention.

Dumas industrialized his pillage by lowering a large basket. The basket plunged down through a hole in the deck and fouled on jumbled beams. Didi went after it, entangled his regulator in the cable, and hung there, unable to move for fear of cutting his air hoses. I happened to be passing and disengaged him. He immediately went down to wrestle with the basket. He filled it and tugged on the line to have his surface helper haul away. The basket crashed back into the ship again. On the next try the compass tripod, which he had lashed on beside the basket, lodged between two rails. Didi freed the compass and the basket fell back into the hold. A strangled cry rang through the sea.

Didi rescued the basket again. He swung it over the side of the ship and it went up very nicely. When Didi unloaded the dishware, which had survived a violent shipwreck and a quarter of a century in the sea, he found it shattered to bits. An unfeeling shipmate said, "Better postpone the wedding, eh, Didi?"

Dumas' affection for the *Dalton* all but ended in tragedy. One day the mistral was blowing too hard to risk a boat, but Didi wanted to finish a film sequence in the aftcastle, so he dived into the thrashing surf with a camera. He went down alone, tossed in the heavy combers. Six feet under, the water was calm and sleepy, but he could feel the crests running overhead by increased pressure on his eardrums. He could sense the passing waves down to twenty feet in the still water. Deep into the lonely quiet, knowing that none of us would follow him down, went Didi, timorous of the least incident, feeling delicate and vulnerable.

His course was the customary one of entering the engine-room hatch and swimming down the big tunnel to the gap we had trespassed, and attaining the aftcastle, which we returned to with pride like small boys who have reached the topmost crotch of a tree.

In the engine room Dumas felt something holding him back by the left air hose, the inhalation pipe. The mask materially reduces vision at the sides, like horse blinders. Dumas could not see the obstacle. He tried to turn his head but the obstruction prevented it. Whatever it was, his hose was straining against it. Didi reached his hand behind his head and felt a pipe covered with dog's teeth clams. He took a razor nick in his hand.

Then before him he saw a pipe that extended toward his head and passed over his left shoulder. It bristled with clams. The pipe passed between his air hose and his neck. Somehow he had looped his hose over a broken end of it and had passed on down. He was held like a quoit on a stake. Miraculously the clams had not cut the hose or his neck tendons. He had no way of knowing how far he had passed down the pipe.

He dropped the camera and hung, without moving a muscle, and thanked Providence that there was no current running in the *Dalton*. He was a hundred feet down, cut off from his comrades by a wild surf, knowing that none of us intended to join him.

Dumas reached back and placed both hands around the pipe to keep it from touching his neck and hose. He drew himself backward, inch by inch, taking new hand grips on the clams, prepared to cut his palms to ribbons to get off the pipe. For an infinite time he recoiled, watching the pipe slowly pass his mask, a handbreadth at a time.

His hand felt the torn end of the pipe and he was free. Dumas's longest undersea journey was ten feet. Without bothering with his hands, he picked up the camera. He ran down the tunnel and filmed a last view of the unearthly cockpit in the supernatural light below the storm. With his work completed, he returned to the lighthouse stairs and thrust his mask out of the brine. A wave swung him up on the stone stairs and he plodded up to the lighthouse.

After Dumas's ordeal on the pipe, we made a rule never to go down alone. It was the beginning of team diving, the essence of aqualung work.

The sea allows each sunken ship to have a personality which is vividly expressed to a diver. In their surface lives the wrecks have had tragic or comic, dull or adventurous careers. We like to check the history of the ships before they entered the depths, and sometimes have been able to uncover lurid peccadillos in the past of a somnolent old *épave*. Such was the case of the *Dalton*, a wild Christmas revel, and another was the adventure of the hot-tempered *Ramon Membru*, a Spanish freighter that lies east of Cavalaire on the Côte d'Azur.

A gray-haired farmer told us the story in a café in Cavalaire. We had hunted for days for this man who had seen the *Ramon Membru* go down in 1925. "It is dawn, *messieurs*," he began, "and I am sitting on the rocks of Cap Lardier with my fishing pole. I perceive an astonishing sight, an enormous ship coming straight at me. To see a big ship this near shore is very odd, but to see it arriving on top of one is not believable. The *Ramon Membru* strikes the reef with a horrible rumbling. It drives fiercely at the reef. The bow climbs up on land and the hull bends like a mass of jelly. There she remains." After twenty years the witness still trembled with excitement as he thought of it.

"All day the Spaniards load launches with suitcases and trunks and take them to the beach. The customs officer of Cavalaire revolts. He declares that if they land any more contraband, he will affix seals on the cargo, which is a shipload of Spanish cigars.

"The next day an ocean-going tug arrives. It very gently removes the ship by the stern. The *Ramon Membru* is afloat — a miracle, *messieurs*. The tug passes a cable to the bow of the *Ramon Membru* to get the ship under tow. The cable parts!

"The coast is very near and a fresh wind pushes the ship again toward the rocks. The tug sees there is no time to waste and succeeds in fastening another cable. The *Ramon Membru* is towed into Cavalaire.

"That night the village is aroused from sleep. The Spanish ship is blazing in the harbor! All the cigars are alight. The *Ramon Membru* goes down."

We found the *Ramon Membru* a few hundred yards from the town jetty, in unclear water which was a rich emerald color. We were surprised to find a vessel of five to six thousand tons. Persons who tell us about sunken ships are given to exaggeration, particularly citizens of the South of France. But our farmer had been a sailor and he spoke the truth. The ship was flattened into the weeds with only the stern and forequarters standing up in relief. Around her was a curious sort of moat in the sand. We found nothing inside the ship, not even the ghost of a cigar band.

But there we met *les liches*. The liche is a pelagic fish the size of a man. It is related to the tuna, but is slimmer and more graceful. It is never taken on lines and it escapes from nets. One must spend some time in the sea before one can see a liche. A noble sight it is, a great silver streamliner moving in regal freedom. Liches came Indian file across the bare plain and passed close to the cigar ship, as though following a primeval trade route. One day they would be hurried and nervous, the next day relaxed and playful. Their comings and goings were unpredictable. Days would pass without sight of the herd, then they would file by again, like a caravan of the desert.

Under our water glass at Port-Cros lay a tiny fishing trawler, a clean wreck newly received by the sea, its nets neatly folded on deck and cork floats tugging at the panels. We did not profane the innocent boat, but the net gave us an idea. We would film a commercial trawl net moving across the floor. No one had seen a trawl net in action. Fishermen who spent their lives dragging the bottom knew the net only in theory. If one found the right spot it brought up fish: that was the existing body of information on trawl nets.

Perched above the grassy floor I saw the towline of the net arrive. It looped back to the rigid gate which scraped along the bottom, breaking down grasses and spreading destruction to the tiny creatures of the prairie. Fishes leaped away like rabbits running from a reaper. The vast envelope of net passed me, puffed up with water. The broken grass arose slowly in the track of destruction. I was astonished to see how many fish escaped the monster, and how much it destroyed of future fish stocks and pasturage. Man's method of undersea farming seemed to consist of blighting the acre while reaping a small part of the crop. Didi hung head-down on the towrope and filmed into the dragon's mouth to bring up evidence on how many fish got away and how much of the nursery was being ruined.

There were bigger nets to see, the web of antisubmarine defenses, closing off the entry to Hyères. Early in the war the main gate had been tended by the deep-sea tug *Polyphème*. Past her prime, she was appointed to guard the door like the aged concierge of a Paris apartment house. At night the *Polyphème* closed the door and anchored and went to sleep with the key in her hand. She was nodding there the night of November 27, 1942, when the fleet exploded at Toulon. The *Polyphème* committed suicide and went down, still moored to the net.

We visited her a year afterward. She lay in sixty feet of exceptionally clear water with her truck, the tip of the mainmast, only four feet down. When we put our masks under we became dizzy. Her hundred and fifty foot bulk was naked and the shrouds and masts, looming intact, took away the sense of its being a wreck. A slight starboard list added to the realistic

Above, floating up the rigging of the sunken tug, Polyphème, *I pan my camera with Dumas as he swims across the weather deck amidst shoals of fish. Below, sunken ships like the tangled skeleton of the Spanish collier* Ferrando, *100 feet down in the Bay of Hyères, snag generations of fishing nets. We have learned to be careful of air hoses in such jungles of steel and gill netting.*

effect. There was not a weed on her yet, merely a fuzz that did not hide the paint.

There was absolutely nothing below decks. The crew had stripped her with fiendish competence before they opened the seacocks. *Polyphème* was a tidy suicide, as bare as an apartment on the last look before the moving men leave.

On the charts of the Bay of Hyères there is a tiny circle marked *épave*, the obituary of the six-thousand-ton Spanish cargo ship, *Ferrando*, sunk fifty years ago. It is precisely drawn on the map, but finding a wreck under the drawing table is another matter. A local bayman took us out in his launch and lost his confidence when he arrived on the spot. "I'm not sure," he said. "It's around here somewhere. . . ." We saw a barrel buoy moored five hundred yards off. The guide had never seen it before. "Perhaps fishermen have placed it to mark where they lost a net," he said.

Dumas went a hundred feet down the buoy line and arrived at the grave of the *Ferrando*. The ship was almost a skeleton and it was festooned with generations of fishing nets. The ship lay on her port flank, so that her remaining deck plates stood like artillery-shattered walls.

Dumas swam into the main cargo hatch, big and dark as a cathedral portal. Inside blue light fell from the starboard portholes and corroded openings and, away off in the hull, a big shaft of light entered through the breach cut long before by the dangling helmet divers who had plundered the *Ferrando*.

Dumas swam down the nave and found, on the invading sand in the afterhold, four china dishes veined in black. Scattered around in the great hull was a rubble of gray-green stones, as ugly as a person sees in a fever. The stones rose in frozen cascades around the cargo hatches. Didi picked up a field-gray stone and smashed it against a bulkhead. It crumbled away in shining black fragments. It was the *Ferrando*'s cargo, bituminous coal with the patina of fifty years in the ocean.

Outside the wreck he looked across the desert at black headstones, the upright shells of the huge pinna mussel. The nets on the *Ferrando* were the grillwork of a cemetery plot, in which were buried the hopes of fishermen. They know that the most fish are to be found near a wreck, and they know that wrecks seize nets. If one drags just so close, there is a good haul, too close and one loses all.

Didi swam through the pinna shells. A hundred yards from the propeller he came upon a small amphitheater in the sand. In the center lay a diminutive Japanese porcelain sake bowl, dainty as an eggshell and undamaged. He put it in his musette bag and continued across a litter of naval shell cases fallen from target practice, and found a cheap earthenware bowl. Its time in the sea had added an expensive-looking crackled finish. He put it in his bag.

The diving table governing his visit was drawing to a close if he was to surface without lengthy stage decompression. As he started up he saw a mathematically straight sand road carved across the bottom. He stopped to examine it. The road ran true into the mists in either direction. Who or what made the road? Where did it go?

Didi came up with his two little dishes. We went back next day to look at the mysterious road, but the barrel buoy was gone. We made dive after dive to find the *Ferrando*. We could not locate the wreck. Didi has the sake bowl and the crackled dish in his new house at Sanary, and a visitor who asks about them receives an interrogation on what he might know about Roman roads on the bottom of the sea.

Chapter Four **Undersea Research Group**

WHEN the German occupation ended I was detailed to Marseilles to run a collecting center for returning sailors in a commandeered castle. One night I tried to assess my past and future: my job was useful but it seemed to me that any officer could do it. On the other hand, our diving experiences, which we had begun with no formal naval directive, now seemed pertinent to the Navy: there were hundreds of jobs for divers in the scuttled fleet and in ships torpedoed at sea.

To convince the Marine Ministry that I was more useful as a diver I went to Paris and showed Admiral André Lemonnier and his staff the film of Dumas and Tailliez finning through the wrecks. The next day I was on my way to Toulon with a commission to resume diving experiments.

Tailliez was happy to leave his makeshift job as a forest ranger. We enlisted Dumas as a civilian specialist, and acquired a desk in the harbormaster's office with a sign, *Groupe de Recherches Sous-Marines* (Undersea Research Group). Philippe, who was senior to me, became commandant. Our table of basic equipment consisted of two aqualungs. Of course, we neglected no opportunity to make ourselves known as a powerful bureau of the *Marine Nationale*.

Three petty officers joined us: Maurice Fargues, Jean Pinard and Guy Morandière. Dumas put them through a rapid aqualung course and made them diving instructors. We promoted funds, men, motorcycles and trucks and, soon, a small craft, a newly built launch named *L'Esquillade*.

After *L'Esquillade* we took over the *VP 8*, a seventy-two-foot twin-screw launch, which Tailliez transformed into a diving tender, with compressed-air supply, diving platform, and decompression chamber. We traded some aqualungs for the Royal Navy's latest frogman suits. Sir Robert H. Davis, the inventor of the Davis submarine escape lung, and head of the world's largest diving and salvage equipment firm, asked for the rights to manufacture the aqualung in Britain.

Our biggest equipment *coup* was the *Albatross*, a real ocean-going diving tender given to us by the Ministry of Marine. The

When the Allies landed in Normandy in 1944 I left Paris by bike to join my family in the Alps, 500 miles away. I carried 110 pounds of food, wine, and documents and made it in four days, pushing up mountain trails to avoid German and maquis *skirmishes.*

Albatross was two years old, but she had a history of hard knocks. Before she was launched or painted, she was seized in a German yard by the Soviets. She was then awarded to Britain and moored in the Thames. A further shuffle of booty sent her to France. The ship had suffered from lack of paint, general neglect and two years of pilferage, while the victors shunted her about. She found a happy berth with the Undersea Research Group. We forgot duty hours as we rigged our ship, which we christened *L'Ingénieur Élie Monnier*, after a naval engineer I had known who had perished in a diving accident.

The sunken S.S. *Dalton* had made deepsea divers of us, and the *Élie Monnier* turned us to oceanography. The ship carried us on voyages to Corsica, Sardinia, Tunisia, Morocco and the open Atlantic. Scientists sailed with us, widening our knowledge of the sea, and themselves attracted to the lung as an instrument of direct observation.

The Navy authorized diving research projects which were conducted by the keen technical men who had joined us, including Dr. F. Devilla, Group surgeon, and pharmacist's mate Dufau-Cazenabe. Jean Alinat was in charge of our "toy shop," where he designed and built new masks, insulated suits, weapons and underwater lighting equipment. There we built the undersea sled, on which a diver could be towed underwater at six knots; it more than tripled the horizontal diving range of search missions.

We introduced a useful feature to be carried on the sledder's belt—a tiny buoy attached to a weighted fishing line on a small reel. When the sled was on reconnaissance, the pilot threw out a buoy when he saw an object of interest and continued his ride. Afterward a diver could follow the buoy line down to investigate the sight.

The Group established liaison with oceanographic and diving establishments in Britain, Germany, Sweden and Italy. The Royal Navy, during the war, had made some valuable studies in the resistance of divers to underwater explosions. Professor J. B. S. Haldane who worked on some of the investigations had written, "Now you must be a very brave man indeed to hunt for magnetic mines in muddy water, especially if you have seen some of your comrades go up. You must be a man of super-human courage if you know that, if you hear a mine beginning to tick, and go up to the surface quickly, you may be paralyzed for life if you are not blown to pieces." Dumas got us into this unappetizing game, because of his incorrigible curiosity about odd effects under water.

One Sunday at Sanary he threw an Italian hand grenade into the water to kill some fish. A few bogues floated up on their sides. Tailliez dived and brought up ten times as many fish he had found lying on the bottom, proof that dynamiting is an excessively wasteful method of fishing unless one can dive to recover the full catch.

Dumas tossed another grenade. It did not explode. Didi waited several minutes and went in to inquire. The grenade lay

We built an undersea sled, pulled by a boat. The pilot glides at six knots, using fin and rudder controls to fly up, down, and sideways, like a towed aerial glider. The sled is used to make rapid reconnaissance of the sea floor.

in fifteen feet of water. As he submerged, Dumas saw tiny bubbles fizzing from it. He did not comprehend what was taking place. The grenade went off directly below him, the worst tangent of an underwater explosion, which blasts vertically through the lighter water layers.

Dumas was in no danger of being hit by shrapnel, which dies out in a dribble a few feet from the burst, such is the resistance of water to flying objects. Death comes from shock waves slamming against the body. Yet Tailliez saw Dumas stagger out of the surf. He was racked to his bones, but actually uninjured. The striking incident set off brain waves. We looked at the British tables on underwater explosions and found that Dumas had been in a radius supposed to cause certain death. We postulated that the shock resistance of a naked man was several times greater than the most advanced study indicated.

Dumas and Tailliez thereupon set out to test the hypothesis. Haldane had said, "Experiments on animals are useful to show the kind of danger to be expected, but they do not tell exactly what a man can stand. This can only be done on human beings whose courage or curiosity will keep them going until they drop."

We went under water in pairs while one-pound TNT charges were exploded at progressively nearer distances. When a burst caused too much discomfort, we stopped. The danger factor was that one might sustain internal lesions without feeling them. The program was concluded without internal injuries, however.

The dynamite explosions boxed our ears disagreeably and dealt a sort of dry smack against the body. With one-pound tablets of German tolite explosive, the effect was different. We took heavy sandbag blows on the chest, receiving a profound shaking. The astonishing thing was how close we could get and bear it. Sometimes, when two of us were hanging in our stations waiting for the explosion, we would look at each other and wince at the lunatic idea.

A third type of explosive I cannot mention gave Tailliez and

Dumas such a bad time that we decided we had better things to do. There had been instructive results. Our first conclusion was that a naked man had better resistance to explosion than a helmet diver. This apparent paradox arises from the fact that pressure waves have nearly the same propagation speed in human tissue as in water, another affirmation of the physical likeness of flesh and the sea mother. For the helmet diver his casque was the weakness. Although it protected his head from the blow, shock waves striking the flexible suit would shudder in his flesh and through his collar and meet no counter pressure inside the helmet.

It is unpleasant to go to war after the war is over, a duty that falls to the minesweepers and demolition teams who grope in the water for lethal objects mislaid by the belligerents. Mine recovery was not one of the paramount aims of the Undersea Research Group, but headquarters have a persuasive way of putting new problems up to naval units.

Salvage and reconstruction in the Toulon yards were hampered by the presence in neighboring waters of unrecovered German mines. Navigation was endangered in the Porquerolles island area, menaced by an unknown distribution of classical contact mines. We began in the usual manner by questioning local fishermen. They produced a chart showing the safe lane they used to sail in and out of Porquerolles. We departed in *L'Esquillade* and ran at full speed down the safe lane, feeling that we were well started on the work. Suddenly in the fading evening light I saw the spiked antennae of a mine passing our side by inches. I slowed down and posted lookouts. The bow lookout called out mine after mine. The "safe" channel appeared to be the heaviest alignment of mines.

Seen from below, mine shells were quite impressive. Barnacles and weeds had appropriated and assimilated them as the sea adopts and transforms any man-made object. The trigger antennae extended toward the surface like monstrous sea-urchin spines; the buoy lines melted into the depths incrusted with mussels. It seemed on our first impression that the mines had lost their deadly character. But they were real and ready, despite the sea change. We had no time charges to blow the mines, but were obliged to arm them with electric wires carried back to our borrowed LCVP, which stood two hundred yards off.

The Navy handed us one of our worst problems in Toulon where a red buoy marked a suspected wreck just off the main fairway. The hulk was shallow enough to endanger shipping. The demolition people were ordered to blow all obstacles out of the way, but a prudent officer saw some strange objects through the surface and held up the dynamiters until we could survey the wreck. Dumas, Tailliez and I swam down and found a big barge loaded high with metal cylinders, overgrown with a light blanket of flora, at which fish were nibbling. We swam around the cargo, examining it closely and scraping away the weeds to affirm that the metal was aluminum. I was photographing the wreck when Dumas seized my arm and drew me rapidly to the surface. "I recognize them now," said Dumas, "they must be those German mines that go off acoustically, magnetically and by pressure waves." Our explosives expert identified the cargo as some of the most fiendish mines the Nazis had evolved, in the late stages of the war. He estimated that there were twenty tons of high explosive down there waiting, a charge sufficient to wreck part of the naval yard and remove most of the leaves, window glass, neon signs, chimney pots and roof tiles in Toulon.

The demolition people studied our photos and stated that an attempt to lift one mine would set off the lot, and that to drag the barge further out would invite the same consequence. Nothing could be done except to stake off a wide area roundabout the wreck and wait for the potent sea to corrode the mechanisms. We left the conference, thinking about the fish nuzzling the mines and the way we had touched them.

At another time I was called in by Captain Bourrague, in charge of a de-mining project for the Ministry of Reconstruction. "I have some cheap katymines for you," he stated genially. "Around Sète in the Gulf of Lions they left a lot of trashy explosives packed in concrete blocks, with the firing mechanisms sticking up on a tripod." He sketched how the trip handles were interconnected from one mine to another by wires so that one mine could explode its neighbors. The poverty-stricken authors of the katymines had also implemented them with cables rising to small buoys anchored off just below the surface to foul ships' screws and trigger the mines.

Minesweepers had canvassed the area three years before, but it seemed that a barge used by the municipality of Sète had encountered one of the miserable things and had gone up in the air.

We went to Sète for some reconnaissance dives. The enemy souvenirs were strewn on an uneven rocky shelf, varying in depth from six to forty feet of cold, turbid water. The concrete blocks had settled comfortably in the mud between the rocks, and in some cases only the tips of the trip handles could be seen. The maze of interconnecting cables had largely deteriorated in six years. The katymines lay along the coast for dozens of miles, often with poor visibility, at most twenty-five feet. I estimated that, if we used customary aqualung search techniques, it would take us many years to be certain that we had located all mines in the total seven and one half acres of the various minefields. The idea of years out of my diving life spent mucking for miserable concrete blocks in cold and dirty water had no appeal. I was driven to inventing a new de-mining technique. I made certain requests of Captain Bourrague for refitting the *VP 8*.

When our ship was ready at Sète the citizenry was highly diverted. We had a towering new mainmast from which guy wires supported two sweeping fifty-five-foot spars on either side. The ensemble was aflutter with varicolored pennants. Five halyards, bedizened with rags a yard apart, ran from the deck through blocks on the spars, and into the water. The halyards were spaced twenty-two feet apart. The center halyard ran out over the stern. Two dories were towed from the ends of the spars. In one was a sailor with a load of tiny white buoys, while in the second was another equipped with a boathook.

The halyards terminated in forty-pound lead weights shaped like fish, to which five aqualung divers clung like subway straphangers. We trolled slowly back and forth along the coast day after day with banners flying and officers crying orders from the bridge to lower and raise halyards. The sailor in the port dory chucked out his little buoys in a line, the ship turned sharply and retraced a parallel course, and the sailor on the starboard dory reaped the buoys with his hook. For two months the *VP 8* plied the water.

We blew fourteen katymines, convinced that we had tagged every last one. The divers were, of course, looking for mines. They were so spaced that they could scan every foot of the floor. When they found a mine they sent up a yellow buoy. The pennants on the halyards marked each yard of depth and the bridge officers hauled in and paid out the divers according to constant scrutiny of the hydrographic chart, so that the divers were trimmed three feet above the changing contour of the floor. We turned half-hour legs so that cold divers could be relieved on each turn.

We were proud of our carnival ship. Only two of the twenty straphangers had previous diving experience. The other volunteers were lads who had checked out in two weeks. The novices were an incessant concern to us, lest they tangle with the explosive wire network, but they all came back as proud menfish.

On the day of the explosions they were sent down free to arm the katymines they had marked with the yellow buoys. They regarded it as a holiday.

At this period the Undersea Research Group developed submarine photography as a survey tool. We had been asked to film "anti-asdic pills," a task that developed into the first motion picture of a submarine navigating under the sea.

Asdic is the undersea equivalent of radar—ultra-sound gear used by the Allies for tracking U-boats. The Nazi counter-device was the anti-asdic pill, a tin box released from the hunted U-boat. The box was ballasted to hang in midwater, where it emitted screens of bubbles which gave the hunters a convincing echo of a U-boat, to lure them into depth-bombing the bubbles. A series of pills from a single submarine boat could convey the impression that a dangerous pack was roaming below.

Philippe Tailliez plunged with a camera and positioned himself on the course of a submarine. It was to come by at moderate periscope depth, discharging anti-asdic pills. Tailliez saw the net cutter on its prow emerge from the haze and pass. In a third of a minute he saw the last visible part fading into the fog, the twin propellers, turning slowly like members of a cheap wind-up toy. The anti-asdic pills behaved according to theory, but Philippe surfaced, wildly impressed with the submarine. It was incomparably more wonderful than fake submarine shots in war movies. We resolved to film a submarine doing all her tricks under the sea. We secured the services of the *Rubis*, a mine-laying submarine.

The day we went down to play with the *Rubis* the water was fairly clear, visibility ninety feet. Lieutenant Jean Ricoul, commanding the submarine, obliged us with a bottoming exercise. We lolled beside the motionless *Rubis*, one hundred and twenty feet down, looking at a periscope blinded to the world outside, a compass dead in its gimbals, a useless machine gun, a radar scope which could not read water, all manned by curious fish. The petrified flag stood unfurled but unwaving, its red and blue panels the same shade of green.

But forty men were cheerfully alive inside the armored air bubble—we heard loud domestic clatter, footfalls, the roaring of a pump, a wrench hitting the deck, almost the curses of the man who dropped it. Then the heavy form stirred up and the screws turned, flattening the weeds and blowing up algae and mud. The *Rubis* climbed. The bow cleaved the surface first. The flag disappeared and the conning tower broke through. The hull was outlined by a dazzling wreath of foam. The mysterious intruder became a prosaic keel.

Ricoul bottomed again for a sequence showing the firing of

a torpedo fifty feet down. Dumas got up on the hull with a hammer in his hand, while I paced off my camera angle. Thirty feet out I sighted into the torpedo tube and moved six feet aside of the probable trajectory. I waved to Didi and he smote the hull with his hammer. The tube hatch flew open and the torpedo bounded out toward me. I had to keep the torpedo framed in my lens as it hurtled past my nose, and to continue tracking it. I convulsed all my muscular power on the grips of the camera to force it around in the water for a panoramic shot. The torpedo roared past like a racing car, and I held on it as it dissolved, smearing a white wakeline through the blue.

Our next game with the *Rubis* was filming an escape exercise from the outside. Ricoul took Didi and Guy Morandière aboard as escapees and settled the *Rubis* one hundred and twenty feet

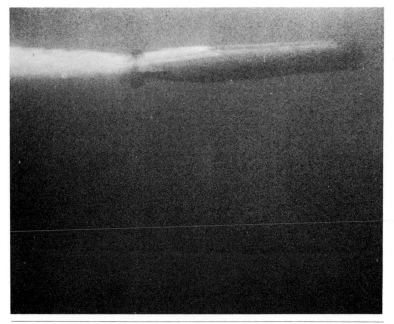

Probably the first movies of a full-sized torpedo fired by a submarine boat. It roared past me, 50 feet down.

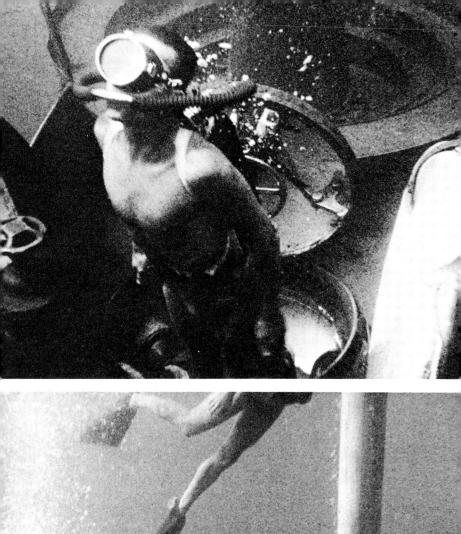

down on the bottom. Didi and Guy clambered into the escape hatch and a submariner locked them in. The divers were sealed in a steel tube, two feet in diameter and seven feet high, an unaccommodating metal bottle. Inside they controlled the valves that flooded the lock and permitted the air to escape. Cold sea water climbed their naked bodies, until the lock was tightly filled. The water pressure rose to where it equalized the pressure outside, almost five times that of the interior of the submarine. It was painful for the eardrums and nerve-racking to the men in the lock, waiting for the escape hatch to open automatically when the pressure was right.

The hatch quivered and opened like an oyster. It liberated a gallon of exhaled air which leaped upward in a happy blob, and the eager escapees slithered up to freedom.

We expected some trouble with our sequence of a mine being laid on the sea floor by a moving submarine. We had strung buoys to form a gangway one hundred and fifty feet wide, down the center of which the *Rubis* would cruise at periscope depth and lay a series of four contact mines on the floor fifty feet down. The mines were the old-fashioned spiked ball type, carried in heavy metal cases, or *crapauds*. A salt tablet in the case would melt in twenty to thirty minutes, thereby releasing the mine, which would ascend on its cable to a point just below the surface where it would be moored off by a hydrostat on its stealthy wait for ships.

The problem was how the *Rubis* was to drop several *crapauds* right into my camera range, so I could film the entire action of a naval mine. We conferred with Lieutenant Ricoul. Dumas said, "I have a plan for making sure the mines fall in front of the camera. I'll ride the sub down and give the order to drop them at the right place." Ricoul and I raised our eyebrows. Didi added, "You do not understand. I am going down

Escape exercise. The submarine boat Rubis *is 120 feet down. Two men at a time are flooded in the airlock until pressure reaches five atmospheres, equal to outside pressure. They open the hatch and swim languidly toward the sun.*

on the *outside* of the submarine." Ricoul diplomatically remained silent, smiling skeptically. "Shall we let him try?" I said.

Ricoul watched Didi straddle the prow with a hammer in one hand and his other clutching the *Rubis*'s netcutter. The skipper went below and applied his eye to the periscope. He didn't want to miss the sight of what was going to happen to the crazy man when the bow crashed down into the furious turmoil of water. He saw the bow waves assault Dumas, throwing his rubber fins up from the spray. Then the rider went under, still hanging on, with his bubbles streaming back past the conning tower. Ricoul wondered how Dumas could hold on against a five-knot avalanche of water, but he was doing it. Ricoul trimmed off at periscope depth and lined up on the gangway.

I felt lost on the bottom, waiting in the cold, keeping alert in my vapory station in the center of the gangway. I had no compass and there is no sense of direction under the sea. I turned constantly, watching for the *Rubis*. Long before she came into view I heard her engine's roar. The sea delivered the muffled growls of the submerged plant from all directions. My apprehension increased as the noise grew louder.

Dumas saw me first; ninety feet out he spotted my bubbles. Then I saw him as the bow pushed into the room. He was a queer little figurine worn like a lucky charm on the gigantic snout. He hammered on the hull and my anxiety vanished at the loud signal to the minelayers inside. How that whale passed! The flanks reeled through, Didi was borne away, and the first *crapaud* fell ten feet from me without a sound and threw up a curtain of mud.

At a rhythm of one every twenty seconds the other mine cases fell out of *Rubis*. Swimming as fast as I could I filmed the arrival of three *crapauds* and stopped for breath as the fourth crashed out of sight. Mud clouds expanded upward from the

The sub lays a contact mine, which rises on its cable toward the surface. Overleaf, a diver of the Undersea Research Group swims through the torpedo hole in an unknown ship that went down off Marseilles in World War II.

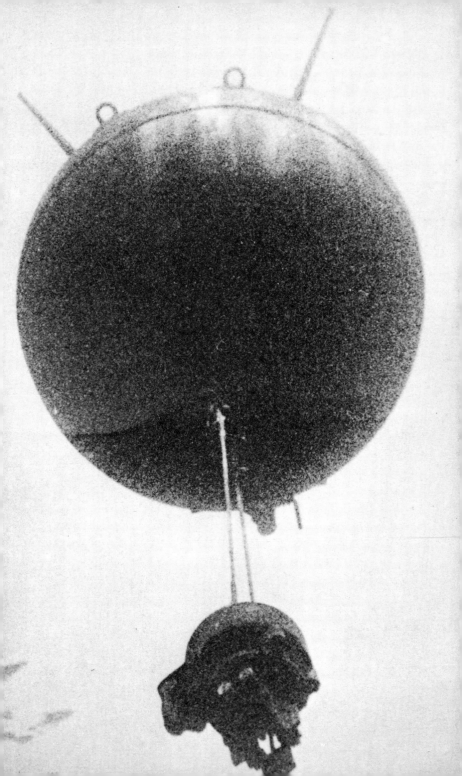

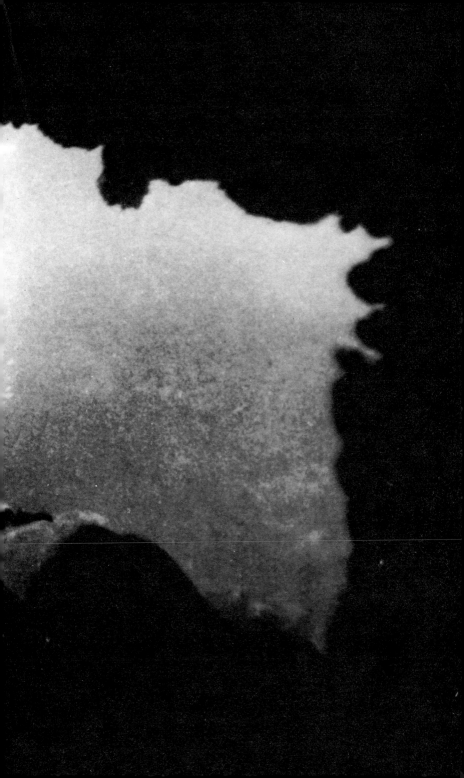

mines, thinned out, and finally revealed the large metal boxes, inside which were the horned mines sucking on their salt tablets. I filmed them from several angles and went back to the surface. Tailliez went down with the camera to pick out a mine and ready himself to film the release.

In five minutes Tailliez heard a distant mechanical rattle. He fingered the camera button. His subject did not stir. One after the other, he heard the other mines go up. Tailliez peered at his choice as long as he could stand the cold. He surfaced after thirty-five minutes to pass the camera. Fargues grabbed it and went down. He was back in a moment. "The bloody ball went up before I got there," he said. "It picked the instant nobody was there."

We appealed to Ricoul to come in and drop four more mines. I said, "We have a good sequence of the *crapauds* landing, so it will not be necessary to carry Frédéric Dumas this trip." The Commandant laughed. He went to sea and came down the gangway again. Didi, the retired pilot fish of the iron shark, was on camera. He prepared to dive. "Pick the right one this time," we yelled. Didi held up his fist, confidently, turned up his flukes and pelted down. He was gone a long time. We looked at each other knowingly and guessed what had happened. The mines were still trying to fool us. We organized a rotation of divers to go down and relieve the camera on the spot, so that it would always be attended. The first man went down and took over from Dumas, who surfaced and said, "The other three went up, but my bastard won't release." Twice we had picked the wrong number out of four.

The divers relieved each other every ten minutes. We were firm with the reluctant mine. We gave the impression that delay was useless. We were going to have a man down there until it came out in surrender. At last the *crapaud* grumbled and the spiked ball went aloft. It had taken one hour. The whole mine sequence lasted ninety seconds on the screen.

After five years the Undersea Research Group had a compact headquarters in a three-story building in the Navy yard, overlooking the berths of its two diving tenders. On the ground floor were the compressor room, experimental compression chambers, machine shop, photo lab, electric converter plants, garage and stalls for experimental animals. The second floor accommodated the drafting room, stowage and crew quarters. Topside were offices, physiological and physics laboratories, chemistry lab and a conference chamber decorated with some of our treasures of the sea: ship's bells and wheels, an Ionic capital and Greek amphoras saved from wrecks. Rising through the three stories was a synthetic diving well, which could simulate pressure of eight hundred feet down.

Other navies had larger diving centers. Ours had the virtue of an intimate connection with the sea beneath the windows. Whether the specialist inside the building was a draftsman or medical orderly, he was also a diver; and the shops and labs were able quickly to exert land techniques on the study of the sea. If Dumas or Alinat imagined a new undersea device, the draftsmen and machinists would have the model ready for tests the next day.

Where there were diving accidents, naval or civilian, the emergency patients were brought to our doctors. Bent and agonized men were carried into the physiological lab for expert decompression in the caisson, and our staff would gather to see the patient emerge, leaping and rejoicing. The orderlies would ask him if he wanted his crutch, which was invariably answered by a happy oath and permission to throw it as far as possible into the ocean.

Among the tasks given to the Undersea Research Group I recall one particularly painful one, the recovery of drowned airmen.

On a misty Mediterranean summer's day a twin-engined naval aircraft was deluded by a mirage and hit at full speed on the hidden swell. The pilot, a friend of mine, was thrown out and picked up dead by a fishing boat. The flying machine skimmed and foundered. We were asked to recover the co-pilot and the engineer.

Their grave was marked by an iridescent tracery of gasoline rising from the wing tanks. Dumas's search dive found the airplane bright on the puce floor one hundred and twenty feet down. It lay upright with the propellers torn off, the engine cowlings ripped open and a wide gash in the cockpit, through which my friend had been projected from the pilot's seat. Small glittering fish thronged around the plane.

I swam among them with a camera, filming for the accident *critique*, and passed the co-pilot seated inside what had been the cockpit, with his eyes wide open. It was the first body I had seen in a thousand dives. I could not bear the calm wise expression of his face. His parachute had mushroomed from the back leg.

A hundred feet away on the spare bottom gorse lay the second flier on his back with one leg flexed and his right forefinger pointing at the surface. His parachute was fully released. It spread like a chute laid out for repacking in the rigging shop. The engineer had probably been catapulted from the vehicle when it hit and the blow had released his parachute. Dumas and Morandière attached lines from the launch and the airmen with their parachutes blossoming below them were raised to the surface.

A twin-engined naval aircraft has crashed and sunk 120 feet down. We find the pilot 30 yards from the plane, encircled by his parachute. Above, a member of the Undersea Research Group has dived 20 fathoms for the airman, and prepares to bring him up. Below, the body is raised to the surface.

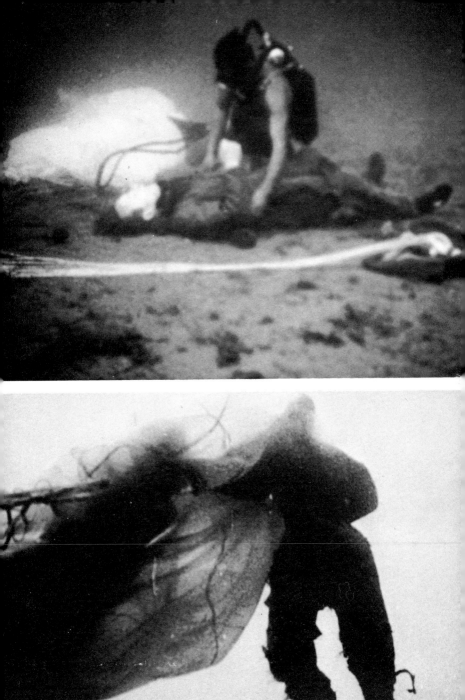

Chapter Five **Cave Diving**

O UR worst experience in five thousand dives did not come in the sea but in an inland water cave, the famous Fountain of Vaucluse near Avignon. The renowned spring is a quiet pool in a crater under a six-hundred-foot limestone cliff above the River Sorgue. A trickle flows from it the year around, until March comes; then the Fountain of Vaucluse erupts in a rage of water which swells the Sorgue to flood. It pumps furiously for five weeks, then subsides. The phenomenon has occurred every year of recorded history.

The fountain has evoked the fancy of poets since the Middle Ages. Petrarch wrote sonnets to Laura by the Fountain of Vaucluse in the fourteenth century. Frédéric Mistral, our Provençal poet, was another admirer of the spring. Generations of hydrologists have leaned over the fountain, evolving dozens of theories. They have measured the rainfall on the plateau above, mapped the potholes in it, analyzed the water, and determined that it is an invariable fifty-five degrees Fahrenheit the year round. But no one knew what happened to discharge the amazing flood.

One principle of intermittent natural fountains is that of an underground siphon, which taps a pool of water lying higher inside the hill than the water level of the surface pool. Simple overflows of the inner pool by heavy rain seeping through the porous limestone did not explain Vaucluse, because it did not entirely respond to rainfall. There was either a huge inner reservoir or a series of inner caverns and a system of siphons. Scientific theories had no more validity than Mistral's explanation: "One day the fairy of the Fountain changed herself into a beautiful maiden and took an old strolling minstrel by the

The Fountain of Vaucluse, near Avignon, France, an age-old mystery that almost took our lives. Here is the entrance, with its high-water mark, across which grows a little fig tree that drinks once a year in the unexplained spring flood that wells out of the depths.

hand and led him down through Vaucluse's waters to an underground prairie, where seven huge diamonds plugged seven holes. 'See these diamonds?' said the fairy. 'When I lift the seventh, the fountain rises to the roots of the fig tree that drinks only once a year.'" Mistral's theory, as a matter of fact, possessed one more piece of tangible evidence than the scientific guesses. There is a rachitic hundred-year-old fig tree hooked on the vertical wall at the waterline of the annual flood. Its roots are watered but once a year.

A retired Army officer, Commandant Brunet, who had settled in the nearby village of Apt, became an addict of the Fountain as had Petrarch six hundred years before. The Commandant suggested that the Undersea Research Group dive into the Fountain and learn the secret of the mechanism. In 1946 the Navy gave us permission to try. We journeyed to Vaucluse on the 24th of August, when the spring was quiescent. There seemed to be no point in entering a violent flood, if its source might be discovered when the fountain was quiet.

The arrival of uniformed naval officers and sailors in trucks loaded with diving equipment, set off a commotion in Vaucluse. We were overwhelmed by boys, vying for the privilege of carrying our air cylinders, portable decompression chamber, aqualungs and diving dresses up the wooded trail to the Fountain. Half the town, led by Mayor Garcin, dropped work and accompanied us. They told us about the formidable dive into the Fountain made by Señor Negri in 1936. What a bold type was this Señor Negri! He had descended in a diving suit with a microphone inside the helmet through which he broadcast a running account of his incredible rigors as he plunged one hundred and twenty feet to the inferior elbow of a siphon. Our friends of Vaucluse recalled with a thrill the dramatic moment when the voice from the depths announced that Señor Negri had found Ottonelli's zinc boat!

We knew about Negri and Ottonelli, the two men who had preceded us into the Fountain, Ottonelli in 1878. We greatly admired Ottonelli's dive in the primitive equipment of his era.

We were somewhat mystified by Señor Negri, a salvage contractor of Marseilles, who had avoided seeing us on several occasions when we sought firsthand information on the topography of the Fountain. We had read his diving report, but we felt deprived of the details he might have given us personally.

The helmet divers described certain features to be found in the Fountain. Ottonelli's report stated that he had alighted on the bottom of a basin forty-five feet down and reached a depth of ninety feet in a sloping tunnel under a huge triangular stone. During the dive his zinc boat had capsized in the pool and slid down through the shaft. Negri said he had gone to one hundred and twenty feet, to the elbow of a siphon leading uphill, and found the zinc boat. The corrosion-proof metal had, of course, survived sixty years of immersion. Negri reported he could proceed no further because his air pipe was dragging against a great boulder, precariously balanced on a pivot. The slightest move might have toppled the rock and pinned him down to a gruesome death.

We had predicated our tactical planning on the physical features described by the pioneer divers. Dumas and I were to form the first *cordée*—we used the mountain climber's term because we were to be tied together by a thirty-foot cord attached to our belts. Negri's measurements determined the length of our guide rope—four hundred feet—and the weights we carried on our belts, which were unusually heavy to allow us to penetrate the tunnel he had described and to plant ourselves against currents inside the siphon.

What we could not know until we had gone inside the Fountain was that Negri was overimaginative. The topography of the cavern was completely unlike his description. Señor Negri's dramatic broadcast was probably delivered just out of sight of the watchers, about fifty feet down. Dumas and I all but gave our lives to learn that Ottonelli's zinc boat never existed. That misinformation was not all of the burden we carried into the Fountain: the new air compressor with which we filled the breathing cylinders had prepared a fantastic fate for us.

We adjusted our eyes to the gloom of the crater. Mayor Garcin had lent us a Canadian canoe, which was floated over the throat of the Fountain, to anchor the guide rope. There was a heavy pig-iron weight on the end of the rope, which we wanted lowered beforehand as far as it would go down. The underwater entry was partially blocked by a huge stone buttress, but we managed to lower the pig iron fifty-five feet. Chief Petty Officer Jean Pinard volunteered to dive without a protective suit to attempt to roll the pig iron down as far as it was possible. Pinard returned lobster-red with cold and reported he had shoved the weight down to ninety feet. He did not suspect that he had been down further than Negri.

I donned my constant-volume diving dress over long woolens, under the eyes of an appreciative audience perched around the rocky lip of the crater. My wife was among them, not liking this venture at all. Dumas wore an Italian Navy frogman outfit. We were loaded like donkeys. Each wore a three-cylinder lung, rubber foot fins, heavy dagger and two large waterproof flashlights, one in hand and one on the belt. Over my left arm was coiled three hundred feet of line in three pieces. Dumas carried an emergency micro-aqualung on his belt, a depth gauge and a *piolet*, the Alpinist's ice axe. There were rock slopes to be negotiated: with our heavy ballast we might need the *piolet*.

The surface commander was the late Lieutenant Maurice Fargues, our resourceful equipment officer. He was to keep his hand on the guide line as we transported the pig iron down with us. The guide rope was our only communication with the surface. We had memorized a signal code. One tug from below requested Fargues to tighten the rope to clear snags. Three tugs means pay out more line. Six tugs was the emergency signal for Fargues to haul us up as quickly as possible.

When the *cordée* reached Negri's siphon, we planned to station the pig iron, and attach to it one of the lengths of rope I carried over my arm. As we climbed on into the siphon, I would unreel this line behind me. We believed that our goal would be found past Negri's teetering rock, up a long sloping arm of the siphon, in an air cave, where in some manner unknown the annual outburst of Vaucluse was launched.

Embarrassed by our pendant gadgetry and requiring the support of our comrades, we waded into the pool. We looked around for the last time. I saw the reassuring silhouette of Fargues and the crowd jutting around the amphitheater. In their forefront was a young *abbé*, who had come no doubt to be of service in a certain eventuality.

As we submerged, the water liberated us from weight. We stayed motionless in the pool for a minute to test our ballast and communications system. Under my flexible helmet I had a special mouthpiece which allowed me to articulate under water. Dumas had no speaking facility, but could answer me with nods and gestures.

I turned face down and plunged through the dark door. I rapidly passed the buttress into the shaft, unworried about Dumas's keeping pace on the thirty-foot cord at my waist. He can outswim me any time. Our dive was a trial run: we were the first *cordée* of a series. We intended to waste no time on details of topography but proceed directly to the pig iron and take it on to the elbow of Negri's siphon, from which we would quickly take up a new thread into the secret of the Fountain. In retrospect I can also find that my subconscious mechanism was anxious to conclude the first dive as soon as possible.

I glanced back and saw Didi gliding easily through the door against a faint green haze. The sky was no longer our business. We belonged now to a world where no light had ever struck. I could not see my flashlight beam beneath me in the frightening dark—the water had no suspended motes to reflect light. A disc of light blinked on and off in the darkness, when my flashlight beam hit rock. I went head down with tigerish speed, sinking by my overballast, unmindful of Dumas. Suddenly I was held by the belt and stones rattled past me. Heavier borne than I, Dumas was trying to brake his fall with his feet. His suit was filling with water. Big limestone blocks came loose and

*The Fountain of Vaucluse, inside the mouth. An arrow points down
through the clear water to the dark door we entered.*

rumbled down around me. A stone bounced off my shoulder. I remotely realized I should try to think. I could not think.

Ninety feet down I found the pig iron standing on a ledge. It did not appear in the torch beam as an object from the world above, but as something germane to this place. Dimly I recalled that I must do something about the pig iron. I shoved it down the slope. It roared down with Dumas's stones. During this blurred effort I did not notice that I lost the lines coiled on my arm. I did not know that I had failed to give Fargues three tugs on the line to pay out the weight. I had forgotten Fargues, and everything behind. The tunnel broke into a sharper decline. I circled my right hand continuously, playing the torch in spirals on the clean and polished walls. I was traveling at two knots. I was in the Paris subway. I met nobody. There was nobody in the Métro, not a single rock bass. No fish at all.

At that time of year our ears are well trained to pressure after a summer's diving. *Why did my ears ache so?* Something was happening. The light no longer ran around the tunnel walls. The beam spread on a flat bottom, covered with pebbles. It was earth, not rock, the detritus of the chasm. I could find no walls. I was on the floor of a vast drowned cave. I found the pig iron, but no zinc boat, no siphon and no teetering rock. My head ached. I was drained of initiative.

I returned to our purpose, to learn the geography of the immensity that had no visible roof or walls, but rolled away down at a forty-five-degree incline. I could not surface without searching the ceiling for the hole that led up to the inner cavern of our theory.

I was attached to something, I remembered. The flashlight picked out a rope which curled off to a strange form floating supine above the pebbles. Dumas hung there in his cumbersome equipment, holding his torch like a ridiculous glowworm. Only his arms were moving. He was sleepily trying to tie his *piolet* to the pig-iron line. His black frogman suit was filling with water. He struggled weakly to inflate it with compressed air. I swam to him and looked at his depth gauge. It read one hundred and fifty feet. The dial was flooded. We were deeper than that. We were at least two hundred feet down, four hundred feet away from the surface at the bottom of a crooked slanting tunnel.

We had rapture of the depths, but not the familiar drunkenness. We felt heavy and anxious, instead of exuberant. Dumas was stricken worse than I. This is what I thought: *I shouldn't feel this way in this depth. . . . I can't go back until I learn where we are. Why don't I feel a current? The pig-iron line is our only way home. What if we lose it? Where is the rope I had on my arm?* I was able in that instant to recall that I had lost the line somewhere above. I took Dumas's hand and closed it around the guide line. "Stay here," I shouted. "I'll find the shaft." Dumas understood me to mean I had no air and needed the safety aqualung. I sent the beam of the flashlight around in search of the roof of the cave. I found no ceiling.

Dumas was passing under heavy narcosis. He thought I was the one in danger. He fumbled to release the emergency lung. As he tugged hopelessly at his belt, he scudded across the drowned shingle and abandoned the guide line to the surface. The rope dissolved in the dark. I was swimming above, mulishly seeking for a wall or a ceiling, when I felt his weight tugging me back like a drifting anchor, restraining my search.

Above us somewhere were seventy fathoms of tunnel and crumbling rock. My weakened brain found the power to conjure up our fate. When our air ran out we would grope along the ceiling and suffocate in dulled agony. I shook off this thought and swam down to the ebbing glow of Dumas's flashlight.

He had lost the better part of his consciousness. When I touched him, he grabbed my wrist with awful strength and hauled me toward him for a final experience of life, an embrace that would take me with him. I twisted out of his hold and backed off. I examined Dumas with the torch. I saw his protruded eyes rolling inside the mask.

The cave was quiet between my gasping breaths. I marshaled all my remaining brain power to consider the situation. Fortunately there was no current to carry Dumas away from the pig iron. If there had been the least current we would have been lost. *The pig iron must be near.* I looked for that rusted metal block, more precious than gold. And suddenly there was the stolid and reassuring pig iron. Its line flew away into the dark, toward the hope of life.

In his stupor, Didi lost control of his jaws and his mouthpiece slipped from his teeth. He swallowed water and took some in his lungs before he somehow got the grip back into his mouth. Now, with the guide line beckoning, I realized that I could not swim to the surface, carrying the inert Dumas, who weighed at least twenty-five pounds in his waterlogged suit. I was in a state of exhaustion from the mysterious effect of the cave. We had not exercised strenuously, yet Dumas was helpless and I was becoming idiotic.

I would climb the rope, dragging Dumas with me. I grasped the pig-iron rope and started up, hand over hand, with Dumas drifting below, along the smooth vertical rock.

My first three hand holds on the line were interpreted correctly by Fargues as the signal to pay out more rope. He did so, with a will. I regarded with utter dismay the phenomena of the rope slackening and made superhuman efforts to climb it. Fargues smartly fed me rope when he felt my traction. It took an eternal minute for me to form the tactic that I should continue to haul down rope, until the end of it came into Fargues's hand. He would never let that go. I hauled rope in dull glee.

Four hundred feet of rope passed through my hands and curled into the cavern. And a knot came into my hands. Fargues was giving us more rope to penetrate the ultimate gallery of Vaucluse. He had efficiently tied on another length to encourage us to pass deeper.

I dropped the rope like an enemy. I would have to climb the tunnel slope like an Alpinist. Foot by foot I climbed the fingerholds of rock, stopping when I lost my respiratory rhythm by exertion and was near to fainting. I drove myself on, and felt that I was making progress. I reached for a good hand hold, standing on the tips of my fins. The crag eluded my fingers and I was dragged down by the weight of Dumas.

The shock turned my mind to the rope again and I received a last-minute remembrance of our signals: six tugs meant pull everything up. I grabbed the line and jerked it, confident that I could count to six. The line was slacked and snagged on obstacles in the four hundred feet to Maurice Fargues. *Fargues, do you not understand my situation?* I was at the end of my strength. Dumas was hanging on me.

Why doesn't Dumas understand how bad he is for me? Dumas, you will die, anyway. Maybe you are already gone. Didi, I hate to do it, but you are dead and you will not let me live. Go away, Didi. I reached for my belt dagger and prepared to cut the cord to Dumas.

Even in my incompetence there was something that held the knife in its holster. *Before I cut you off, Didi, I will try again to reach Fargues.* I took the line and repeated the distress signal, again and again. *Didi, I am doing all a man can do. I am dying too.*

On shore, Fargues stood in perplexed concentration. The first *cordée* had not been down for the full period of the plan, but the strange pattern of our signals disturbed him. His hard but sensitive hand on the rope had felt no clear signals since the episode a few minutes back when suddenly we wanted lots of rope. He had given it to us, eagerly adding another length. *They must have found something tremendous down there*, thought Fargues. He was eager to penetrate the mystery himself on a later dive. Yet he was uneasy about the lifelessness of the rope in the last few minutes. He frowned and fingered the rope like a pulse, and waited.

Up from the lag of rope, four hundred feet across the friction of rocks, and through the surface, a faint vibration tickled Fargues's finger. He reacted by standing and grumbling, half to himself, half to the cave watchers, *"Qu'est-ce que je risque? De me faire engueuler?"* (What do I risk? A bawling out?) With a set face he hauled the pig iron in.

I felt the rope tighten. I jerked my hand off the dagger and hung on. Dumas's air cylinders rang on the rocks as we were borne swiftly up. A hundred feet above I saw a faint triangle of green light, where hope lay. In less than a minute Fargues pulled us out into the pool and leaped in the water after the senseless Dumas. Tailliez and Pinard waded in after me. I gathered what strength I had left to control my emotions, not to break down. I managed to walk out of the pool. Dumas lay on his stomach and vomited. Our friends stripped off our rubber suits. I warmed myself around a flaming caldron of gasoline. Fargues and the doctor worked over Dumas. In five minutes he was on his feet, standing by the fire. I handed him a bottle of brandy. He took a drink and said, "I'm going down again." I wondered where Simone was.

The Mayor said, "When your air bubbles stopped coming to the surface, your wife ran down the hill. She said she could not stand it." Poor Simone had raced to a café in Vaucluse and ordered the most powerful spirit in the house. A rumor-monger raced through the village, yelling that one of the divers was drowned. Simone cried, "Which one? What color was his mask?"

"Red," said the harbinger.

Simone gasped with relief—my mask was blue. Then she thought of Didi of the red mask and joy collapsed. She returned distractedly up the trail to the Fountain. There stood Didi, a miracle to her.

Dumas's recuperative powers put the color back on him and his mind cleared. He wanted to know why we had been

---

Above, the moment of our escape from death in the Fountain of Vaucluse. At top of picture, I am surfacing, half-gone, but holding a line to the still-submerged Dumas (lower in the picture), who passed out 400 feet down in the crooked galleries, at an actual depth of 200 feet. Below, Dumas is dragged out.

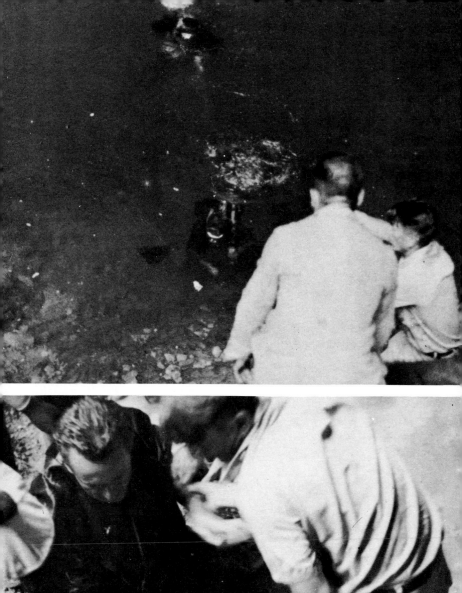

drugged in the cavern. In the afternoon another *cordée*, Tailliez and Guy Morandière, prepared to dive, without the junk we had carried. They wore only long underwear and light ballast, which rendered them slightly buoyant. They planned to go to the cavern and reconnoiter for the passage which led to the secret of Vaucluse. Having found it they would immediately return and sketch the layout for the third *cordée*, which would make the final plunge.

From the diving logs of Captain Tailliez and Morandière, I am able to recount their experience, which was almost as appalling as ours. Certainly it took greater courage than ours to enter the Fountain from which we had been luckily saved. In their familiarization period just under the surface of the pool, Morandière felt intense cold. They entered the tunnel abreast, roped together. Second *cordée* tactics were to swim down side by side along the ceiling.

When they encountered humps sticking down from the roof, they were to duck under and return to follow closely the ceiling contour. Each hump they met promised to level off beyond, but never did. They went down and down. Our only depth gauge had been ruined, but the veteran Tailliez had a sharp physiological sense of depth. At an estimated one hundred and twenty feet he halted the march so they might study their subjective sensations. Tailliez felt the first inviting throbs of rapture of the depths. He knew that to be impossible at a mere twenty fathoms. However, the symptoms were pronounced.

He hooted to Morandière that they should turn back. Morandière maneuvered himself and the rope to facilitate Tailliez's turnabout. As he did so, he heard that Tailliez's respiratory rhythm was disorderly, and faced his partner so that Tailliez could see him give six pulls on the pig-iron rope. Unable to exchange words under water, the team had to depend on errant flashlight beams and understanding, to accomplish the turn. Morandière stationed himself below Tailliez to conduct the Captain to the surface. Tailliez construed these activities to mean that Morandière was in trouble. Both men were slipping into the blank rapture that had almost finished the first *cordée*.

Tailliez carefully climbed the guide line. The rope behind drifted aimlessly in the water and a loop hung around his shoulders. Tailliez felt he had to sever the rope before it entangled him. He whipped out his dagger and cut it away. Morandière, swimming freely below him, was afraid his mate was passing out. The confused second *cordée* ascended to the green hall light of the Fountain. Morandière closed in, took Tailliez's feet and gave him a strong boost through the narrow door. The effort upset Morandière's breathing cycle.

We saw Tailliez emerge in his white underwear, Morandière following through the underwater door. Tailliez broke the surface, found a footing and walked out of the water, erect and wild-eyed. In his right hand he held his dagger, upside down. His fingers were bitten to the bone by the blade and blood flowed down his sodden woolens. He did not feel it.

We resolved to call it a day with a shallow plunge to map the entrance of the Fountain. We made sure that Didi, in his anger against the cave, could not slip down to the drowned cavern that had nearly been our tomb. Fargues lashed a one hundred and fifty foot line to Dumas's waist and took Didi's dagger so he couldn't cut himself loose and go down further. The final reconnaissance of the entrance shaft passed without incident.

It was an emotional day. That evening in Vaucluse the first and second *cordées* made a subjective comparison of cognac narcosis and rapture of the Fountain. None of us could relax, thinking of the enigmatic stupor that had overtaken us. We knew the berserk intoxication of *l'ivresse des grandes profondeurs* at two hundred and twenty feet in the sea, but why did this clear, lifeless limestone water cheat a man's mind in a different way?

Simone, Didi and I drove back to Toulon that night, thinking hard, despite fatigue and headache. Long silences were spaced by occasional suggestions. Didi said, "Narcotic effects aren't the only cause of diving accidents. There are social and subjective fears, the air you breathe . . ." I jumped at the idea. "The air you breathe!" I said. "Let's run a lab test on the air left in the lungs."

The next morning we sampled the cylinders. The analysis showed 1/2000 of carbon monoxide. At a depth of one hundred and sixty feet the effect of carbon monoxide is sixfold. The amount we were breathing may kill a man in twenty minutes. We started our new Diesel-powered free-piston air compressor. We saw the compressor sucking in its own exhaust fumes. We had all been breathing lethal doses of carbon monoxide.

Further expeditions were made to the caves of Chartreux and Estramar which taught us much about the problems of cave diving. But we still had not gone through a siphon or the mechanism that shot water earthward. In 1948, while most of us were away on the *Bathyscaphe* expedition, three members of the Group finally achieved the goal, Lieutenant Jean Alinat, Dr. F. Devilla and CPO Jean Pinard, this time assisted by the Army Corps of Engineers. The spring of Vitarelles near Gramat was the object of their large cave expedition.

Vitarelles is a subterranean spring. The surface of the water is three hundred and ninety feet down. The engineers carried out a full-scale dry-cave operation before the divers reached the water. First the soldiers descended an air shaft two hundred and seventy feet deep, lowering pontoons, duckboards, aqualungs, constant-volume suits, lines, electric-lighting equipment, and food. From this landing they conveyed the equipment down another hole, narrow and almost vertical, one hundred and twenty feet to an underground chamber. From this base they were required to lay duckboards and pack the gear sixteen hundred feet through partially flooded galleries, including a dangerous cramped passage thirty feet long. Only then did they reach the surface of the spring, into which the divers were to continue for hundreds of feet more. The engineers established a pontoon pier in the pool, with diving ladders, and the sailors prepared to dive.

I prepare to go into the Fountain of Estramar, inland near Per-pignan, France. My jacket is the foam-rubber "mid-season" dress Dumas designed for chilly work.

Alinat's plan was to send divers down one at a time, on safety ropes of progressively greater lengths. Using measuring lines, flashlights, compasses, depth gauges and sketch blocks, the divers mapped the water tunnel, each one advancing further than the man before. The scheme worked smoothly, and the chart moved league-by-league into the void. The culminating tenth dive was made by Alinat on the 29th of October, 1948.

The diver before him had reached the entrance to a siphon. Alinat went down, fastened to a four-hundred-foot safety line, and rapidly swam to the limit of the chart. The gallery rose at a twenty-degree angle. Alinat swam into the narrow tunnel. He passed uphill through forty feet of rather turbid water in a darkness pierced only by his narrow flashlight beam. He felt his head part a gentle tissue and water resistance ceased. Through his mask, now blurred like a windshield in rain, he saw that his head was in air. He was in a sealed clay vault one hundred and fifty feet long. He removed his mouthpiece and mask and breathed natural air. Where water flows, even in a sealed pocket beneath the earth, there is air.

He climbed out on a slippery strand that ranged down one side of the long room. He was the first living thing in the vault of water, earth and air, where no sun had ever brought the gift of life. He walked along the shore, measuring and sketching, elated with the victory of our campaigns against the fountains.

At the far end, Alinat received a bitter revelation. Plain under the clear water was the aperture of another siphon. The mechanism of Vitarelles held further secrets. Alinat sat down and thought of the cost of penetrating the new labyrinth. The divers would have to transport equipment nearly four hundred feet under water to set up an advanced camp in the clay room, before they could plunge into the second siphon.

Alinat finished his sketch and walked back to the entrance, imprinting rubber frog tracks on the hidden beach. He spit in the mask and sloshed it in the water. He molded the mask over his face and inserted the mouth grip. He slipped into the water,

A diver goes down through submerged weeds in darkness and a rising current.

turned up his flukes and sailed head down through the current of the first siphon. In a few minutes his exhalations sputtered out on the surface. Nothingness was restored in the cave. The tracks of man vanished into darkness.

Chapter Six **Treasure Below**

I~N~ A Hyères waterfront bistro a fisherman's story of a wreck brought us to the edge of our chairs. "A long time ago," he said, "two paddle-wheelers collided between Ribaud and Porquerolles Islands and sank very deep. One ship contained gold. You ask Michel Mavropointis, the old Greek diver. He worked on them. A salvage contractor tried to pump the mud away to get at the wreck. The government lent two warships to turn it over, but the vessel resisted. It lies keel up. Nobody can penetrate it. No divers will go near it since one of them met a monstrous conger eel in the funnel."

It was the customary wreck tale, an upside-down ship with an accessible funnel and a standard sea monster guarding the gold. But Michel Mavropointis was a friend of ours and it seemed worth investigation. Dumas and I repaired to Michel's favorite bar. The retired diver was overjoyed to see us. He could always depend on us to hear his tales out. To us old Michel represented the honorable profession of diving.

He accepted a *pastis*, poured it into a glass of water, and quaffed the cloudy drink. "The two ships you seek have nothing in common," he stated. "One of them is the *Michel Say*, a big three-hundred-footer sunk fifty years ago. The other is the *Ville de Grasse*, a paddle-wheeler that was cut in two by the *Ville de Marseilles*, around 1880. The *De Grasse* carried Italian emigrants. There were fifty-three victims. The emigrants carried one thousand, seven hundred and fifty pieces of gold down with them. The bow lies in one hundred and fifty feet. The stern is in one hundred and eighty."

Michel's treasure account was in the classic form. It had an exact casualty roll, precise inventory of the gold, name of ship, year of sinking, and depth. If these things were left vague no one would believe a gold saga or invest in an expedition to salve the treasure.

Our friend stacked another saucer and carried on. "I got an adjudication from the government to salvage the *Michel Say*. It gave me the right to anything within eight hundred feet of her. While I worked, I kept looking for the *Ville de Grasse*." We saw Michel's point: he could be secretly hunting gold while he was ostensibly diving in a prosaic wreck. He said, "I worked a whole summer without finding the paddle-wheeler. From the *Michel Say* I took fine china, crystal ware, case after case of excellent beer, and sacks of bread flour. . . ."

"Bread flour?" Didi asked.

"Exactement," said Mavropointis. "The sea formed a crust of flour and sacking and left the contents dry." He savored the professional point he had scored over the johnny-come-latelys, and drained his *pastis*. "At the end of the season I was having a snack with my men on the tender. We were drifting without being aware of it. Suddenly our anchor hooked on something. It was the *Ville de Grasse*, the gold ship, *mes enfants*! Believe me, I lost no time throwing myself into the water. I knelt in the wreck and dug into the mud with my hands. My divers stirred up mud for days. Then I found the chest." Michel drank slowly to give his revelation impact.

"There was great emotion aboard the tender," he continued. "I admit the knives pushed themselves out of the pockets, as the crew gathered around the chest. The lid gave open. The chest was heaped with glittering ornaments of the theater, hair combs, frivolities. When we touched them, they vanished into mud. The gold of the *Ville de Grasse* is still down there."

Michel did not mention the location, and ignored our sly efforts to lead him to the point. We thought it was a sign that there actually was gold in the wreck. We resolved to find the emigrant treasure ourselves, by diving in the known location of the *Michel Say*.

Didi found the wreck on his second plunge. He saw a

shadowy mass which became the serrated profile of an old iron wreck. The *Michel Say* was larger in Mavropointis's memory than in life. It was a boneyard of a ship one hundred and sixty feet long, instead of three hundred. The crayon lines of a bridge thinned away above the bulk of the main boiler, and the broken blades of a small screw stuck out on the bare shaft like a turkey duster. Dumas entered the wreck and searched the sand. Sure enough, there was a sack of flour Michel had left behind years before.

In the *Michel Say* the area of the forepeak was thronged with the most numerous fish we have ever seen. The sedentary kinds were arranged by species as if posing for an ichthyologist, and among them fell a cyclone of pelagic fish. From the wreck Didi took a brass ship's lantern and a pair of binoculars joined to a heap of stones. But excursions from the *Michel Say* brought no sighting of the *Ville de Grasse*. We returned again and again to hunt it. The *Michel Say* thinned away before our eyes as the years passed. The *Ville de Grasse* remained as one of Mavropointis's least reliable entertainments, and he himself died without telling where the paddle-wheeler rested.

One day in 1949, however, we were passing the same area in the research ship *Élie Monnier* when the echo-sound tape recorded another massive object on the floor not far from the *Michel Say*. Didi harnessed up and went down like a hungry porpoise.

It was very deep. Dumas arrived on a soft sand prairie. He saw a vague bulk. As he swam toward it, a surpassing sight materialized, two huge green-fleeced paddle-wheels standing upright on the floor. Between them lay a big steam engine. One side of the *Ville de Grasse* was staked off by its naked ribs, protruding from the sand. There was no vestige of the other side. Where it should have been, hundreds of empty cosmetic jars lay on the sand. Among them Didi found a small copper saucepan and some glass flasks. His wrist depth meter showed fifty-five meters (one hundred and eighty-one feet). Old Mavropointis had told the truth; the depth was exactly as he had given it. Didi saw that excavating the ruin was hopeless. His breathing pipe was leaking and he was swallowing a mixture of salt water and compressed air. He picked up a jar and a flask and headed up. On the way the compressed air in his stomach expanded. He returned from the realm of gold with a hard distended stomach and spent three hours belching as profoundly as after a glorious Chinese dinner.

Dumas received an urgent free-lance treasure-diving commission from Auguste Marcellin, the generous guide to our first sunken ships. He said, "The government is adjudicating four freighters off Port Vendre, down by the Spanish border, the *Saumur*, the *St. Lucien*, *L'Alice Robert* and *L'Astrée*. They were torpedoed by a Free French submarine in 1944 as they came in from Spain. My information is that they were carrying a million dollars worth of wolfram to the Germans. I want to bid on the ships, but I want to be sure what's down there. Didi, how about having a look?" Dumas was flattered to be selected over Marcellin's accomplished helmet divers. He packed off immediately with his gray wooden aqualung cases.

At Port Vendre he toured the bars and talked to sailors, fishermen, *fonctionnaires* and barkeepers. For once the citizenry had a practically unanimous version of the fate of ships. The submarine, linked to an Allied intelligence network in Spain, had crept in under the cliff of Cape Bear, just outside the harbor. On the bluff above there were German coastal guns, but the sub was huddled so close that the guns could not be depressed to bear on it. The submarine commander knew exactly what he was doing. He hung there and sank the wolfram ships, one by one, as they plodded into Port Vendre.

Local divers had penetrated *L'Alice Robert* and found her empty. *L'Astrée* had been torpedoed at night, no one knew where. But everyone in Port Vendre, it seemed, had seen the *Saumur* go down in the harbor mouth, after taking two torpedoes. Dumas engaged a veteran harborman, *le Père* Henri, to take him to the location of the *Saumur*. Papa Henri cast anchor and Didi went down the chain into turbid gloom. Didi's report went: "I make out the shape of a mast. It is covered with thousands of mussels. Two fine lobsters balance in the top shroud ring. It is the first time I have ever seen lobsters on a mast. A long way down I find a winch bristling with lobster horns. I go inside through an open hatch and land on the first cargo deck on a heap of empty mussel shells. This does not seem to be wolfram. I go further into the hold, and find no mussel shells. It occurs to me that some fish had eaten the mussels from the mast and dropped the empty shells into the hold.

"The hold is piled with small stones mixed with earth. I have never seen wolfram, but, if this is it, it isn't very beautiful. I chuck some of the stuff into my musette bag and go up on deck. Lobsters are everywhere. I touch the anchor chain and they tumble out of the hawsepipes. It is getting cold so I surface."

Didi submitted his mineral samples to a local chemist, who said, "Is this wolfram? My dear chap, no. Looks like a poor grade of iron ore to me." The next day Didi returned to the *Saumur*, at a depth of one hundred and fifty-five feet and visited the remaining cargo holds. They were full of the same uninteresting matter the chemist had scorned. He fooled around for a bit on the main deck. Where the torpedo blast had removed deck plates, Didi looked down into cabins and saw lobsters crawling in the basins and bathtubs. He flipped down and picked up three lobsters for Papa Henri.

The *Saumur* was improvident. What about the *St. Lucien*? The local history went, "She received a torpedo forward and the bow anchors dropped with a roar of chain. She nosed down and hung for a long time with her propellers grinding in the air. The second torpedo missed and exploded on the rocks. The *St. Lucien* slid under very slowly."

Dumas went out with *le Père* Henri who dragged a grapnel back and forth for hours, unable to strike the wreck, while Didi lounged in the sun applying deduction to the evidence: *the anchors dropped on the first torpedo. Therefore the ship was involuntarily anchored when she disappeared. She went nose down.* Didi suggested to Papa Henri that perhaps the *St. Lucien* had gone down, swinging around in a spiral off her anchors. "Let's drag nearer shore," Didi suggested. The bayman dragged inshore and struck the *St. Lucien* in the same depth as the *Saumur.* Didi pelted down the cable and arrived at the grapnel, which was standing on a bare bottom with no ship in sight. The grapnel had come clear of the wreck. But it had left a furrow on the floor. Didi followed the trail to the ship.

He planed into the main cargo hold. It was empty, except for thousands of thin wooden splints strewn willy-nilly across the deck. Dumas surfaced, bemused with his treasure hunt. In the morning he awoke to a clamor of women in the courtyard of his hotel. He looked down and saw the local Spanish orange market. The women were uncrating oranges from splint boxes — the splints in the hold of the *St. Lucien.* The ship had gone down so reluctantly because she was full of floatable oranges.

Auguste Marcellin did not bid on the "wolfram" ships.

People who have heard of our activities ask three standard questions. The first is, "In all those wrecks what treasures have you found?" This chapter attempts to answer that one.

The second riddle is, "What about the monsters that guard sunken ships?" That can be answered later.

The third goes, "What happened when you found your first human remains?" We have learned to be patient with the question, which arises from myth, because we would like to restore the truth about the sea and sunken ships. We have dived more than five hundred times in about twenty-five sunken ships, into every crawl hole accessible to a man with three metal bottles on his back, and have never found a trace of human remains. Very few victims drown inside sinking ships. They get off beforehand and drown in the sea.

But suppose a person has been unlucky enough to go down inside a ship. One would have to penetrate the vessel within the first few weeks to find any trace of the body. The flesh is eaten in a period of days, not only by fish and crustaceans, but by such unsuspected creatures as the starfish, which is actually a voracious individual. The bones will then be efficiently consumed, mainly by worms and bacteria.

Legends of undersea treasure are ninety-nine per cent hoaxes or swindles, in which the only wealth uncovered is that which passes from the investor to the promoter. The get-rich-quick aberration that resides in most of us is never more successfully exploited than by treasure promoters with faded maps of sunken galleons. They have a large area of credulity in which to operate because the investors have been under the sea even less than the promoters. A serious salvor, well-equipped with knowledge and technique for recovering a treasure, will keep his activities as clandestine as possible. The very fact that public subscriptions are invited for treasure expeditions is almost a guarantee that the attempt is not so much for underwater gold as for surface money.

I can imagine no worse catastrophe for a shipmaster than actually to discover a treasure. First of all, he would be obliged to inform his crew and sign them to the customary salvage contracts, providing shares for all. Then, of course, he would wear them to secrecy. The second drink by the third mate in the first waterfront bar would take care of the secrecy. At that point, should the skipper have exhumed Spanish gold, heirs and assigns of the *conquistadores* and kings would turn up with genealogical claims on the loot. The government of the country whose territorial waters he had plumbed would present an enforceable tax lien. If the poor chap got back home with any pieces of eight, his own government would remove most of it in taxes. I can see him losing his friends, his reputatation and his ship to the sheriff, and wishing he had left the accursed boodle where he found it.

We were cured of gold fever early in the game, although Dumas has recurring bouts, like malaria. There are modern treasures in sunken ships, tons of tin, copper and wolfram awaiting salvage. They are not in the convenient bulk of the Portuguese *moidores* of treasure promoters. They require big serious salvage operations, controlled by owners, governments or insurance companies, and carried out by long, dull efforts, calculated on a narrow profit margin.

The only successful get-rich-quick salvage operation we have encountered was at Do Sal Island in the Cape Verdes. A man came aboard in this untoward place and greeted us familiarly. We recognized an amateur goggle diver from the Riviera and asked, "What are you doing here?" He said, "Salving a wreck." I asked, "What contractor are you with?" "Nobody," he replied. "I'm working alone." I suspected that we were dealing with a case of hallucination. Our friend insisted that he had a salvage contract, for a vessel in twenty-five feet of water. He was working it alone, using only mask and fins. "I brought one of your narghile outfits but I can't get an air pump around here," he stated. The narghile, or Turkish water pipe, is an adaptation of the aqualung, which, instead of using compressed-air bottles on the diver's back, feeds pumped air through a pipe from the surface. I asked him what the treasure was, prepared for the customary tale of gold and silver bars.

"Cocoa beans," said the visitor. "Four thousand tons of cocoa beans. I have the hatch off. I am at work." We had to depart too soon to investigate this affair but in Dakar a representative of a responsible marine insurance company bore the man out. "He is on contract to us. He has no salvage equipment whatsoever. Except for butterfly nets."

"Butterfly nets!" said Didi.

"Exactly," said the insurance man. "The jute bags of cocoa beans are floating against the overheads. He has a native in a small boat anchored over the wreck. The diver goes down holding his breath, swims in under the hatch and cuts open the bags. He pushes the beans toward the hatch. They float to the surface. The native scoops them up with a butterfly net. Already there is a considerable mound of beans on the beach."

One day, a year later, Didi and I encountered the salvor himself on the Côte d'Azur. He exuded signs of affluence. He looked healthy. "My friends," he said, "it was the finest year I have spent. My salvage award was eight million, seven hundred and fifty thousand francs ($25,000)."

This proves that there really is treasure under the sea.

Chapter Seven **The Drowned Museum**

THERE are finer treasures in the Mediterranean, waiting within range of the lung. She is the mother of civilization, the sea girt with the oldest cultures, a museum in sun and spray. The grandest of undersea discoveries, to our taste, are the wrecks of pre-Christian ships on the floor. Twice we have visited classic wrecks and recovered riches beyond gold, the art and artifacts of ancient times. We have located three more such vessels which await salvage.

No cargo ship of antiquity is preserved on land. The Viking ships that have been found buried in the earth and the Emperor Trajan's pleasure barges which were recovered by draining Lake Nemi in Italy, are splendid evidence of noncommercial vessels of ancient times, but little is known of the merchant ships that brought nations together.

My first clue to the classic ships appeared in the Bay of Sanary, where forty years ago a fisherman brought up a bronze figurehead. He died before I came to Sanary and I have never been able to learn where he found it.

Years later Henri Broussard, leader of the Undersea Mountain-Climbing Club of Cannes, came up from an aqualung dive with a Greek amphora. The graceful two-handled earthenware jar was the cargo cask of antiquity, used for wine, oil, water and grain. The cargo ships of Phoenicia, Greece, Carthage and Rome carried thousands of amphoras in racks in the hold. The bottom of the amphora is conical. On land it was punched into the earth. On shipboard it probably fitted in holes in the cargo racks. Broussard reported that he saw a pile of amphoras in sixty feet of water. He did not guess that it indicated a wreck, because the ship was completely buried.

We dived from the *Élie Monnier* and found the amphoras tumbled and sharded on a bed of compacted organic matter in a dusty gray landscape of weeds. With a powerful suction hose we tunneled down to find the ship. A hundred amphoras came out of the shaft, most of them with corks still in place. A few had well-preserved waxen seals bearing the initials of ancient Greek wine merchants.

For several days we siphoned mud and amphoras. Fifteen feet down we struck wood, the deck planking of a freighter, one of two ancient cargo vessels that have been found. We were not equipped to carry out full-scale salvage and our time was limited. We went away with amphoras, specimens of wood, and the knowledge of a unique hydro-archeological site which awaits relatively simple excavation. We believe the hull is preserved and could be raised in one piece. What things that wreck might tell of the shipbuilding and international commerce of the distant past!

Of ancient ships we know a smattering from murals and vase paintings and can make fairly sound guesses at the science of their navigators. Their cargo ships were short and broad and probably could not work to windward. The few existing lighthouses were fires kept burning on shore and there were no beacons or buoys on rocks and shoals. The skippers must have hated to lose sight of land and always tried to moor at night. The pilots must have inherited a knowledge of generations to risk voyaging a ship. Sentenced to skirt the shore, the ships were prey to sudden Mediterranean storms and treacherous rocks. Most of them that foundered, therefore, must have gone down in relatively shallow littoral waters, within diving range. Naval battles and piracy added to the toll of wrecks in shallow depths. I believe there are hundreds of ancient hulks preserved in accessible mud.

A ship that settled in less than sixty feet of water has probably vanished in the scattering action of tide and current, but if it landed deeper it lies in the calm museum of the floor. If the ship fell on rock bed and could not be wholly swallowed, it was overcome by the intense life of the sea. Algae, sponges, hydrozoa, and gorgonians enveloped it. Hungry fauna sought food and shelter in the wreck. Generations of shellfish died and were crunched by other animals that rained excremental sand and mud which mounted as the wreck broke down. After centuries the simultaneous enveloping and consuming actions reached a common level and the sea bottom healed, leaving perhaps a scar.

A diver needs trained eyes to find the signs of such a wreck —a slight anomaly of the bottom contour, an odd-shaped rock, or the graceful curve of a weed-grown amphora. Broussard's amphoras must have been deck cargo. Amphoras in the hold would have been covered with the ship. Many ancient ships have been lost beyond trace when coral or sponge divers, ignorant of the probability that amphoras point to a ship beneath, removed the jars without noting their location.

Unmistakable were the signs of the only other classic cargo vessel ever found, the Galley of Mahdia. The designation is a misnomer; the ship had no tiers of oars; it was a pure sailing vessel which was specially designed to carry an incredible load for its day, at least four hundred tons. The argosy of Mahdia was built by the imperial Romans nearly two thousand years ago for the express task of looting the art treasures of Greece. Our finding of the argosy climaxed an archeological detective story.

In June, 1907, one of the gnomelike Greek divers who roam far and deep in the Mediterranean was prospecting for sponges off Mahdia on the Tunisian east coast, when he found one hundred and twenty-seven feet down row after row of huge cylindrical objects, half buried in the mud. He reported that the bottom was covered with cannons.

Admiral Jean Baehme, in command of the French Tunisian naval district, sent helmet divers to investigate. The objects consisted of sixty-three cannons lying in apparent order in a scattered oval on the sea plain, along with other large rectangular forms. All were heavily encrusted with marine life. The divers raised one of the cylinders. When the organisms were removed, marble fluting was revealed. The "cannons" were Greek Ionic columns.

Alfred Merlin, the government director of antiquities in Tunisia, sent the news to the famous archeologist and art historian, Salomon Reinach. Reinach aroused art patrons to finance a salvage effort. Two Americans subscribed, an expatriate who styled himself the Duke of Loubat, after a Papal patent, and James Hazen Hyde, who gave $20,000. Reinach guaranteed no results, but Hyde was willing to back the effort. The expedition was in charge of a Lieutenant Tavera, who engaged expert civilian divers from Italy and Greece, equipped with the latest helmet diving suits.

The depth was a serious problem at that stage of diving technique. That year the Royal Navy Deep Diving Committee was working out the first tables of stage decompression for operations to one hundred and fifty feet, of which Tavera did not yet have knowledge. Several divers were so heavily stricken with the bends that they could never work again. The difficult and dangerous operation was pursued for five years.

The argosy was a museum of classic sculpture. It held not only capitals, columns, plinths and horizontal members of the Ionic order, but carved *kraters*, or garden vases, as tall as a man. The divers found marble statuary and bronze figures scattered across the floor as though they had been deck cargo, strewn as the ship side-slipped down like a falling leaf.

Merlin, Reinach and other experts attributed the art to Athens of the first century B.C. They believed that the argosy had foundered about 80 B.C., while carrying the systematically gathered loot of the Roman Dictator Lucius Cornelius Sulla, who had sacked Athens in 86 B.C. The evidence was that the

---

Aqualungers of the "Undersea Mountain Climbing Club" bring up an ancient leaden anchor from the Mediterranean floor. Specialists tentatively believe it is Phoenician, on the evidence of the head carved in the metal (detail below). (Photos by V. Romanovsky.)

architectural members constituted a complete prefabricated temple or sumptuous villa which Sulla's art commissioners had taken apart in Athens for shipment to Rome. The ship was way off course for a journey from Greece to Rome, a not uncommon dilemma of the clumsy sailing ships of that era. Enough *objets d'art* were brought up to fill five rooms in the Museum Alaoui in Tunis, where they may be seen today. In 1913 the salvage operation was broken off when financial aid ran out.

We first heard of the argosy in 1948, when we made an undersea archeological investigation of the supposed sunken commercial harbor of ancient Carthage. The summer before, Air General Vernoux, commanding in Tunisia, had personally taken some curious aerial photos of the shallow water off Carthage. Through the clear sea were seen distinct geometrical forms that startlingly resembled the moles and basins of a commercial harbor. The photos were examined by Father Poidebard, a Jesuit scholar, who was also an Air Force chaplain. He found underwater remains of the ports of Tyre and Sidon in the early twenties and was eager to look into the Carthage discovery.

Father Poidebard came aboard the *Élie Monnier* and we took a ten-man diving team to examine the harbor. We found no trace of masonry or man-made construction, and to check our conclusions we had a powerful dredge cut trenches through the "harbor" features. The dredgings held no traces of building material.

Then in the Tunisian archives and in the Alaoui Museum we came upon the story of the argosy of Mahdia. Merlin's monographs and Lieutenant Tavera's report led us to believe there were many treasures still left in the wreck. I had a thrill when I came across the name of Admiral Jean Baehme: he was my wife's grandfather. When we found Tavera's clear detailed sketches, showing the bearings of the wreck, we went for it.

We lay off shore in dazzling Sunday morning sunlight, studying the sketches. There were three drawings of landmarks which could be aligned to bring us over the argosy. The first alignment was a castle sighted past a stone buttress in a ruined jetty. We saw the castle immediately, but there were four piers of the fallen jetty which could be lined up on it.

The second bearing was to bring a small bush on the dunes in line with the crest of a hill. In the thirty-five years since Tavera had drawn the lonely bush, a veritable forest had grown up around it. The last clue was a change in color of a distant olive grove lined up on a foreground windmill. We squinted through the glasses until our eyes wavered but saw no windmill. We made disparaging remarks about Lieutenant Tavera, now a deceased admiral, and wished he had studied treasure-map cartography from Robert Louis Stevenson.

We went ashore to look for the ruins of the mill. We loaded a truck with wooden beams and muslin to construct a signal beacon on the site. Up and down the dusty road we went, questioning the natives. No one remembered the mill, but some-one suggested the old eunuch might know. We found him hobbling down the road, a withered octogenarian with a bald head and fluffy white sideburns. It was difficult to imagine him as he once must have been, the sleek and proud factotum of an Arabian Nights' harem. His blank eyes lighted encourage-ingly. "Windmill? Windmill?" he squeaked. "I'll take you to it." Carrying our gear, we followed him several miles across coun-try to a pile of rubble. We hurried to build the beacon. The ancient looked worried and mumbled to me, "I remember another mill further on." He took us to a second heap of stone. As we regarded it with pain, he thought of still another ruined mill. The coast of Mahdia seemed to be a graveyard of wind-mills.

We returned to the *Élie Monnier* and held council. We de-cided to exert the maximum possibilities of aqualung search technique to rediscover the wreck as though we knew nothing of its location. That was not greatly exaggerating the situation. We had two facts—the wreck was somewhere near and it was in one hundred and twenty-seven feet of water. Echo sound

Philippe Tailliez breaks the surface, removes his mouthpiece, and shouts the thrilling news: he has found the ancient Roman argosy 130 feet down off Tunisia. For a week we have been prowling the sea floor trying to locate the vessel, which went down in the first century B.C. *with a shipload of art treasures probably looted from Greece by the Roman Sulla. (Photo by Marcel Ichac.)*

established that the floor was nearly level with slight variations in depth. We cruised until we found the depth area closest to Tavera's sounding.

On the sea floor we laid a steel wire grid covering one hundred thousand square feet. It was patterned like an American football field more than doubled in size, with fifty feet between each crossline. The object of the undersea game was for divers to swim to and fro along the stripes, surveying the terrain right and left for signs of the wreck. It took us two days to canvass the grid. We would have found a watch dropped on the field. There was no Roman freighter in our web.

Lieutenant Jean Alinat proposed that he go down on the undersea sled. We towed Alinat around the outskirts of the grid. He found nothing. So passed the fifth unfruitful day of the argosy hunt. That night we indicated our desperation by deciding to search closer to shore.

Next morning Commandant Tailliez waved the sled aside and elected to be towed on a shotline by an auxiliary tender. In our campaigns against the unmanageable sea, I believe I felt the low point that morning, the sixth day of failure. I mentally composed a report to my superiors in Toulon which would explain why it was necessary for me to work two naval vessels and thirty men for a week on a wreck that was salved in 1913. Father Poidebard was beginning to remind me of an angry admiral.

A lookout shouted. Out on the sunny water bobbed a tiny dot of orange plastic, the personal signal buoy Tailliez carried on his belt. When the little buoy comes up, the diver has marked something important. Tailliez broke water, tore out his mouth grip, and yelled, "A column! I found a column."

The old records indicated that one pillar had been dragged away from the wreck and abandoned when operations were terminated. The argosy was ours. We ran into Mahdia for the night and broke out champagne for all hands. What occurred in the bistros that night illumines the problem of a crew that has found undersea treasure. The town buzzed with the news that we had found the fabled golden statue of the galley, a mythological object locally venerated for a third of a century. Philippe's mollusk-eaten pillar became a fortune in gold. Admirers thronged aboard to congratulate us.

We began work at daybreak. Dumas I went down and found the main wreck site. It looked nothing like a ship. The fifty-eight remaining columns were vague cylinders covered with thick blankets of vegetation and animals. They lay pounded, flattened into the muddy basement. We called on our imaginations to flash the picture of a ship. She must have been a whale of a vessel in her day. Tape measurements on the distribution of the columns outlined a ship perhaps one hundred and thirty feet long by forty feet wide, twice the displacement of the *Élie Monnier*, hanging in the sky above.

The argosy was lost in a bare prairie of mud and sand that spread beyond sight into the clear depths. It was an oasis for fish. Big rock bass swam in the drowned museum. We noticed there were no commercial varieties of sponges growing on the columns. The thoroughgoing Greek sponge divers of our day had apparently gleaned them all. Perhaps they had also lifted small art objects as a belated patriotic recovery of the Roman pillage.

We were confronted with a semi-industrial salvage operation. We were heirs to the great advances in diving science since Tavera's brave men had dared the wreck, and we had, in fact, a set of unique diving tables newly worked out under Lieutenant Jean Alinat's direction. They were designed for aqualung work, in which men could go down and come back quickly in a series of short dives, without building up the nitrogen saturation of prolonged single plunges. The latest helmet diving tables for a man who was to work forty-five minutes at the depth of the argosy required him to return by stages to decompress. He had to halt four minutes at a depth of thirty feet, proceed to twenty feet and spend twenty-six minutes, and halt for another twenty-six minutes at ten feet, before surfacing.

It cost him almost an hour to return from a three-quarter-hour dive. Alinat's schedule, in contrast, sent a man down for three fifteen-minute dives, alternated by three-hour rest periods. The independent diver needed only five minutes of stage decompression at ten feet after the third dive, one-twelfth of the helmet diver's decompression wait.

To make Alinat's theories work for an efficient attack on the Roman ship, the two-man teams had to go down and come up on a rigid timetable. They could not be expected to consult their own wrist watches. We devised a "shooting clock," a rifleman on deck who fired into the water five minutes after they had gone down, again at ten minutes, and discharged three rounds at fifteen minutes as the imperative signal to surface. The shock impact of bullets could be heard distinctly in the wreck.

On the first day I saw a diver surface, holding up a small glittering object, and my heart leaped, for we had hoped to find Greek bronzes. It was merely a bullet from the shooting clock. The floor became covered with them. It would have been fun to have been hiding behind a column when the next sponge diver sneaked down and saw the bottom gleaming with gold.

The timetable was also threatened by the fact that the *Élie Monnier* swung wide in wind and current of its one anchor, so that the divers had long unpredictable diagonals to swim on their way to work, a drain of time and energy. Dumas hoisted out on deck a slingload of miscellaneous dockyard scrap he had foraged, such as rusted girder bolts and hunks of plate. The divers laughed at the boyish simplicity of Didi's solution. Holding a fifteen-pound scrap iron against his belly, a diver could go sailing down, using his body as a hydrofoil to control his slanting glide. He could come on the wreck from any approach by adjusting his ballast. He could drag, side-slip, or plunge; arrive rested; and drop his iron commutation ticket.

Didi dutifully obeyed the shooting clock, until one day he spied something fascinating as he climbed from his third dive. The sun was still aglow on the floor. Dumas could not resist a lightning dive at it. He found nothing of interest and returned. At dinner he remarked on a twinge in his shoulder. We kidnapped him instantly, locked him in the recompression chamber on deck, and dialed the inside pressure to four atmospheres. We could not take chances on the bends, which can hit a diver some time after he surfaces. There was a phone from the recompression chamber to a loudspeaker in the divers' ready room. After we had eaten Dumas took the mike and broadcast a diatribe against shipmates who starved a pal. We cooled him off for an hour. It was the only time we have used the recompression chamber on our dives.

The world of the argosy was twilight blue in which flesh was a greenish putty color. The far-off sun gleamed on the chromed regulators, winked on the frames of the masks, and ensilvered our exhaust bubbles. The blond basement suffused a reflected light strong enough to make a color movie of the divers at work. I believe it was the first color film made at such a depth.

The Athenian marbles were dark bluish shapes, blurred with blankets of marine life. We dug under with our hands, dog-fashion, to pass cargo slings under them. As the stones ascended, color grew on the crust and, at the surface, they swung into the air ablaze with life. As they drained on deck, the Joseph's coat of flora and fauna faded into the earth shade of death. We scraped, scrubbed and hosed the snowy marble volutes and bared them to their first sun since ancient Athens.

Of the stones on the floor, we took four columns, two capitals and two bases. We raised two mysterious leaden parts of

Above, we hoist aboard an Ionic capital which had been buried in the sea bed since before Christ was born. Lieutenant Jean Alinat fingers marble carving almost as perfect as it came from the Greek sculptor's chisel over two thousand years ago. (Photo by Marcel Ichac.) Below, Father Poidebard, S.J., in the sun helmet, is an authority on classic archeology. Aboard the Navy research ship, Élie Monnier, he surveys a marble column the divers brought up from the Roman argosy. It is covered with corals, sponges, mollusks and worms. The padré dated the marble as first century B.C. (Photo by Marcel Ichac.)

Dumas, wearing his "mid-season" dress of foam rubber, flies along the reef under La Ciotat with red gorgonians he picked in the twilight zone. The first time we ever saw this supple species of coral was on an underwater ramble during the war. At 132 feet the coral looked dark blue. Here, with color-corrected flash-photo technique we show the true colors of the blue zone for the first time.

Above, Dumas passes the time of day with a spider crab in a cave of yellow sponges. Right, he visits a cave of colorful coral.

A diver entering a coral cave must be aware of its appearance in the sea's deceiving color filter. Above, white-suited divers inspect coral in an underwater cave. Left, Dumas looks at a colorful coral reef in the Red Sea.

Above, an Undersea Research Group diver visits the S.S. Donator, *a Greek steamer torpedoed in 1944, now lying 150 feet down.*
Below, a diver inspects another wreck.

The sea allows each sunken ship to have a personality which is vividly expressed to a diver. Above, divers discover a wreck. Below, a diver mans the helm on a voyage to nowhere.

Fish seem to glide forever as long as one does not startle them. Here a school of Grunt fish swims by unperturbed.

*Looking toward the surface from 25 feet down, this shot made in natural
light shows how the sea's blue filter all but washes out the spectrum.*

Top, spotted Serranidae in the Indian Ocean. Bottom, a grouper—the ocean's scholar—peers inquisitively at us through a rock, its large, touching eyes full of puzzlement.

Top, the needlefish is a gay type always trying to leap across the barrier of water and air. Bottom, a school of angelfish swim by a coral reef in the Indian Ocean.

A nosy shark cruises past, watching the surface, which is where he finds his usual meals of distressed or dead animals and ships' garbage.

Sharks are attracted to a diver entering the water (here an aqualunger with a flash-bulb reflector) but become less menacing when a man swims down into the blue.

Opposite page: parasol coral formations. Top, a squirrel fish swims underneath; bottom, two parasols. Above, angel fish.

Above:An instant ago this niche in the Mediterranean under Sanary was the black of millennial time. Algae and sponges grow in the vault. Below: Sixty square feet of undersea jewelry, 12 fathoms down off Cassis—the ceiling of a grotto of semiprecious red coral. The growths resembling caterpillars are the same red coral, covered with live coral-building animals. In the center are yellow sponges.

Above, a Red Sea parasol coral formation. Below, a reef wall 22 fathoms below Cap Sicié. The colors are as loose and rich as the palette of Pierre Bonnard. The little pink damsel-fish flirt past a biological incrustation six feet thick. Dumas has drilled into such reefs and detonated tiny TNT charges to measure the thickness.

The rich diversity of life in tropical waters. Colorful reef fish are super-imposed over a backdrop of vibrantly colored animals encrusting the coral reef: hard reef-building corals; red sponges and ascidions, passively filter-ing food from the waters; and various types of algae, carrying on the vital process of photosynthesis.

ancient anchors, which were found near the supposed outlines of the ship in positions that indicated the anchors had been stowed when the vessel sank. She must have met her fate suddenly. The anchor parts, each weighing three-quarters of a ton, were oblongs with reinforced holes in the middle, obviously to take wooden posts that had rotted away. Such straight metal shapes could not have been the arms or hooks of the anchors. We dug around for the arms and found none. The finds could only have been the stocks, or top cross-bars. The rest of the anchors must have been made of wood. Here was a puzzle. Why did the ancients put the greatest weight at the top of the anchor?

We argued over evidence and supposition and formed a possible explanation. The ancient ships did not have anchor chain, but used rope. A modern anchored ship driven by wind or current keeps its hooks fast by means of the horizontal stress on the lower end of the anchor chain. The Roman anchor rope drew taut in such conditions and would have lifted the wooden hooks if the top had not been weighted with a leaden stock, which provided the horizontal stress.

We worked six days in the Roman argosy, increasingly absorbed with its clues to original seamanship. We wanted to dig for the ship itself. Tavera's records indicated that the helmet divers had excavated extensively in the stern. I selected marbles from a compact area on the starboard side amidships and sent them up to clear an area for excavation. We lowered a powerful water hose to blow away the earth. A slight current conveniently carried off the mud we raised. We supposed that the heavily laden vessel had burst her top framing outward in the crash and that the main deck had been hammered in by the deck cargo. The theory seemed to be borne out.

Two feet down we stubbed our fingers on a solid deck covered with leaden plates. The sea washed mud into the hole almost as fast as we clawed, but we felt enough of the sturdy deck to estimate that the Roman ship is two-thirds intact. We dug up an Ionic capital which was encased entirely in mud. No mollusks or plants had reached it. It scrubbed down to the pristine beauty of the days when it was carved, before Christ was born.

I am confident that amidships there is unbreached cargo. I am certain that then as now the crew lived in the forecastle, the least desirable place of a ship, and that there are intimate possessions and tools buried there that could tell us about what kind of men sailed the Roman ship.

We were merely scratching at history's door in our few days in the huge argosy. We found iron nails corroded to needle thicknesses and bronze nails worn to bright threads. We turned up a millstone, with which the sea cooks had ground grains carried in amphoras. We brought up yard-long pieces of Lebanon cedar ribs covered with the original yellow varnish. (It would be useful to know how to make marine varnish that will survive twenty centuries of immersion.) I dug down five feet at the prow against the sliding sands and embraced the cedar stempost. I could barely touch fingertips around it.

Four years later in New York I met the president of the French Alliances of the United States and Canada, a lively old gentleman named James Hazen Hyde, and linked his name with that of the patron who had helped salve the treasures of Mahdia. It was the same man. He invited me to dinner at the Plaza and I showed him the color film of the divers in the wreck. "Fascinating," he said. "You know I've never seen the things that were brought up. In those days one had a lot of money and a steam yacht. I was cruising in the Aegean while they were diving. I never got to the museum at Tunis. Salomon Reinach sent me photographs of the *kraters* and statues, I got a nice letter from Merlin, and the Bey of Tunis gave me a decoration. It is interesting indeed to see it after forty-five years."

Chapter Eight **Fifty Fathoms Down**

W E CONTINUED to be puzzled with the rapture of the depths and felt that we were challenged to go deeper. Didi's deep dive in 1943 had made us aware of the problem, and the Group had assembled detailed reports on its deep dives. But we had only a literary knowledge of the full effects of *l'ivresse des grandes profondeurs* as it must strike lower down. In the summer of 1947 we set out to make a series of deeper penetrations.

Here I must say that we were not trying for record descents, although the dives did set new world marks. We have always placed a reasonable premium on returning alive. Even Didi, the boldest among us, is not a stunt man. We went lower because that was the only way to learn more about the drunken effect, and to sample individual reactions on what aqualung work could be done in severe depths. The attempts were surrounded with careful preparations and controls, in order to obtain clear data. The objective range we set was three hundred feet or fifty fathoms. No independent diver had yet been deeper than Dumas's two hundred and ten feet.

The dives were measured by a heavy shotline hanging from the *Élie Monnier*. On the line at sixteen-and-one-half-foot intervals (five meters) there were white boards. The divers carried indelible pencils to sign their names on the deepest board they could reach, and to write a sentence describing their sensations.

To save energy and air, the test divers descended the shotline without undue motion, carried down by ten-pound hunks of scrap iron. They retarded their descent by holding the line. When a man reached the target depth, or the maximum distance he could stand, he signed in, jettisoned his weight, and took the line back to the surface. During the return the divers halted at depths of twenty and ten feet for short periods of stage decompression to avoid the bends.

I was in good physical condition for the trial, trained fine by an active spring in the sea, and with responsive ears. I entered the water holding the scrap iron in my left hand. I went down with great rapidity, with my right arm crooked around the shotline. I was oppressively conscious of the Diesel generator rumble of the idle *Élie Monnier* as I wedged my head into mounting pressure. It was high noon in July, but the light soon faded. I dropped through the twilight, alone with the white rope, which stretched before me in a monotonous perspective of blank white signposts.

At two hundred feet I tasted the metallic flavor of compressed nitrogen and was instantaneously and severely struck with rapture. I closed my hand on the rope and stopped. My mind was jammed with conceited thoughts and antic joy. I struggled to fix my brain on reality, to attempt to name the color of the sea about me. A contest took place between navy blue, aquamarine and Prussian blue. The debate would not resolve. The sole fact I could grasp was that there was no roof and no floor in the blue room. The distant purr of the Diesel invaded my mind — it swelled to a giant beat, the rhythm of the world's heart.

I took the pencil and wrote on a board, "Nitrogen has a dirty taste." I had little impression of holding the pencil, childhood nightmares overruled my mind. I was ill in bed, terrorized with the realization that everything in the world was thick. My fingers were sausages. My tongue was a tennis ball. My lips swelled grotesquely on the mouth grip. The air was syrup. The water jelled around me as though I were smothered in aspic.

I hung witless on the rope. Standing aside was a smiling jaunty man, my second self, perfectly self-contained, grinning sardonically at the wretched diver. As the seconds passed the jaunty man installed himself in my command and ordered that I unloose the rope and go on down.

I sank slowly through a period of intense visions.

Around the two hundred and sixty-four foot board the water was suffused with an unearthly glow. I was passing from night to an intimation of dawn. What I saw as sunrise was light reflected from the floor, which had passed unimpeded through the dark transparent strata above. I saw below me the weight at the end of the shotline, hanging twenty feet from the floor. I stopped at the penultimate board and looked down at the last board, five meters away, and marshaled all my resources to evaluate the situation without deluding myself. Then I went to the last board, two hundred and ninety-seven feet down.

The floor was gloomy and barren, save for morbid shells and sea urchins. I was sufficiently in control to remember that in this pressure, ten times that of the surface, any untoward physical effort was extremely dangerous. I filled my lungs slowly and signed the board. I could not write what it felt like fifty fathoms down.

I was the deepest independent diver. In my bisected brain the satisfaction was balanced by satirical self-contempt.

I dropped the scrap iron and bounded like a coiled spring, clearing two boards in the first flight. There, at two hundred and sixty-four feet, the rapture vanished suddenly, inexplicably and entirely. I was light and sharp, one man again, enjoying the lighter air expanding in my lungs. I rose through the twilight zone at high speed, and saw the surface pattern in a blaze of platinum bubbles and dancing prisms. It was impossible not to think of flying to heaven.

However, before heaven there was purgatory. I waited twenty feet down for five minutes of stage decompression, then hurried to ten feet where I spent ten shivering minutes. When they hauled in the shotline I found that some impostor had written my name on the last board.

For a half hour afterward I had a slight pain in the knees and shoulders. Philippe Tailliez went down to the last board, scribbled a silly message, and came up with a two-day headache.

Dumas had to overcome dramas of heavy rapture in the fifty-fathom zone. Our two tough sailors, Fargues and Morandière, said they could have done short easy labor around the bottom. Quartermaster Georges visited the bottom board and was dizzy for an hour or so afterward. Jean Pinard felt out of condition at two hundred and twenty feet, signed in, and sensibly returned. None of us wrote a legible word on the deep board.

In the autumn we undertook another series of deep dives, with marker boards extending below fifty fathoms. We planned to venture beyond with lines tied to the waist, and a safety man stationed on deck, completely equipped to jump in and give aid in case of difficulty.

Diving master Maurice Fargues dived first. On deck we regularly received the reassuring conventional signal Fargues gave by tugging on the line, *"Tout va bien"* (All is well). Suddenly there was no signal. Anxiety struck us all at once. Jean Pinard, his safety man, went down immediately, and we hauled Fargues up toward one hundred and fifty feet, where they would meet. Pinard plunged toward an inert body, and beheld with horror that Fargues's mouthpiece was hanging on his chest.

We worked for twelve hours trying to revive Fargues, but he was dead. Rapture of the depths had stolen his air tube from his mouth and drowned him. When we brought up the shotline we found Maurice Fargues's name written on the three hundred and ninety-six-foot board. Fargues gave his life a hundred feet below our greatest penetrations, deeper than any helmet diver breathing unmixed air has ever gone in the sea.

He had shared our unfolding wonderment of the ocean since the early days of the Research Group; we retain the memory of his prodigal comradeship. Dumas and I owed our lives to Maurice Fargues, who had resurrected us from the death cave

IN MEMORIAM MAURICE FARGUES, Officier des Equipes, Groupe de Recherches Sous-Marines de Toulon. *Our friend and companion of the first years under the sea, skilled diver, commander of the diving ship, VP 8. Fargues's last signature (below) in indelible pencil was found on the marker board attached to the shot line, 396 feet down, where he drowned.*

at Vaucluse. We will not be consoled that we were unable to save him.

The death of Fargues and the lessons of the summer showed that three hundred feet is the extreme boundary of compressed-air diving. Amateurs can be trained in a few days to reach one hundred and thirty feet, and there professionals, observing decompression tables, may do almost any sort of hard work. In the next zone down to two hundred and ten feet experienced divers may perform light labor and make short explorations if rigid safety rules are followed. In the zone of rapture below only the highly skilled aqualunger may venture for a brief reconnaissance. Free divers could range considerably beyond the fifty-fathom layer by breathing oxygen mixed with lighter gases such as helium and hydrogen. While it has been proved that helium removes the causes of depth drunkenness, such dives would still require long tedious hours of decompression.

Dumas slightly improved the standing free-diving record in 1948 on a mission that had no such intent. He was called out to survey an obstacle believed to be an uncharted wreck, that had fouled the drag cables of a minesweeper. When he came aboard he learned that the depth had been sounded at three hundred and six feet. Dumas kicked up his fins and swam down in ninety seconds. The cable was snagged on a low rock. He studied the situation for a minute and returned as quickly as he had descended. He had not been subjected to enough nitrogen saturation to cause the bends.

Dumas planned the diving courses for the fleet aqualung divers, two of whom are to be carried on each French naval vessel. He immerses the novices first in shallow water to bring them through the fetal stage that took us years—that of seeing through the clear window of the mask, experiencing the ease of automatic breathing, and learning that useless motion is the enemy of undersea swimming. On his second dive the trainee descends fifty feet on a rope and returns, getting a sense of pressure change and testing his ears. The instructor startles the class with the third lesson. The students go down with heavy weights and sit on the floor fifty feet down. The teacher removes his mask and passes it around the circle. He molds the mask again, full of water. One strong nasal exhalation blows all the water through the flanges of the mask. Then he bids the novices emulate him. They learn that it is easy to stop off their nasal passages while the mask is off and breathe as usual through the mouthgrip.

A subsequent lesson finds the class convened at the bottom and again their attendance is assured by weights. The professor removes his mask. Then he removes his mouthpiece, throws the breathing tube loop back over his head and unbuckles the aqualung harness. He lays all his diving equipment on the sand and stands naked except for his breechclout. With sure, unhurried gestures he resumes the equipment, blowing his mask and swallowing the cupful of water in the breathing tubes. The demonstration is not difficult for a person who can hold a lungful of air for a half minute.

By this time the scholars realize they are learning by example. They remove their diving equipment entirely, put it back on, and await the praise of the teacher. The next problem is that of removing all equipment and exchanging it among each other. People who do this gain confidence in their ability to live under the sea.

At the end of the course the honor students swim down to a hundred feet, remove all equipment and return to the surface naked. The baccalaureate is an enjoyable rite. As they soar with their original lungful, the air expands progressively in the journey through lessening pressures, issuing a continuous stream of bubbles from puckered lips.

The first foreign naval officer to visit us officially at Toulon was Lieutenant Hodges of the Royal Navy, who became an enthusiastic free diver and cinematographer. In 1950 there befell to him a tragic task—locating the sunken submarine H.M.S. *Truculent*. In a January fog in the Thames estuary the small Swedish tanker *Divina* rammed and sank the submarine with eighty men aboard. Fifteen of the crew escaped with Davis lungs; from their reports it did not seem difficult to locate the wreck. But the river was cold and filthy with a strong current running. Helmet divers made repeated descents and were swung away in the current, without making contact with the hull. Hodges volunteered to dive with the aqualung. He went in upstream at a point calculated in relation to the speed of the current and the supposed location of the *Truculent*, and went sailing giddily down in the turbulent darkness. He hit the submarine on his first pass. Unfortunately the trapped men were dead when the submarine was raised.

During the summer of Liberation I came home from Paris with two miniature aqualungs for my sons, Jean-Michel, then seven, and Philippe, five. The older boy was learning to swim but the younger had only been wading. I was confident that they would take to diving, since one does not need to be a swimmer to go down with the apparatus. The eyes and nose are dry inside the mask, breathing comes automatically and the clumsiest kick will do for locomotion.

We went to the seashore and I delivered a short technical lecture, which the boys did not hear. Without hesitation they accompanied me to a shallow rocky bottom, amidst sea wrack, spiny urchins and bright fish. The peaceful water resounded with screams of delight as they pointed out all the wonders to me. They would not stop talking. Philippe's mouthpiece came loose. I crammed it back in place and jumped to Jean-Michel to restore his breathing tube. They tugged at me and yelled questions as I shuttled between them, shoving the grips back between their teeth. In a short time they absorbed a certain quantity of water, and it was apparent that nothing short of drowning would still their tongues. I seized the waterlogged infants and hauled them out of the water.

I gave another lecture on the theme that the sea was a silent world and that little boys were advised to shut up when visiting it. It took several dives before they learned to hold their volleys of chatter until they had surfaced. Then I took them deeper. They did not hesitate to catch octopi with their hands. On seaside picnics Jean-Michel would go down thirty feet with a kitchen fork and fetch succulent sea urchins. Their mother dives too, but without the same enthusiasm. For reasons of their own, women are suspicious of diving and frown on their menfolk going down. Dumas, who has starred in seven underwater films, has never received a fan letter from a woman.

Chapter Nine **The Submarine Dirigible**

ONE evening in 1948 my wife said, "Please don't go down in that horrible machine. Resign from the Piccard expedition. Everyone is worried about you." I was surprised. It was the first time Simone had objected to any of my plans. She is a Navy wife with a self-disciplined attitude toward my activities. "No one ordered you to go," she continued. "Don't risk yourself in that craziness." My parents added their objections and several scientists warned me against the depth vehicle of Professor Auguste Piccard.

I comforted them, "The *Bathyscaphe* is perfectly safe. There is nothing to worry about." That was stretching it a bit, because there were some undecided aspects of the project. But Tailliez, Dumas and I were together again, about to sail to West Africa on our greatest adventure, and nothing could stop us. I had been selected to enter a wonderful submarine dirigible and dive five times deeper into the sea than man had ever gone. Professor Piccard, who had been eleven miles in the sky, now proposed to go thirteen thousand feet into the ocean abyss.

The elderly scientific extremist had designed the *Bathyscaphe* ("Depth-craft") a decade before and, after the delay of a World War, it had been built by a brilliant Belgian physicist, Dr. Max Cosyns. The Belgian National Scientific Research Fund paid for the men and mother ship for the great dive. The Undersea Research Group had enlisted the French Navy with reconnaissance and rescue aircraft, two frigates and the Group's *Élie Monnier*. Two French scientists were going along, Professor Theodore Monod, director of the Institute of Black Africa, and Dr. Claude Francis-Beouf, founder of the Centre de Recherches et d'Études Océanographiques. I was detailed as "sea expert," more or less security officer of the expedition.

Our Group had spent two years in preparation; we had constructed much of the bizarre auxiliary equipment of the *Bathyscaphe*, including the deadliest undersea weapon ever made. We welded a camera mount on the forefoot of the *Élie Monnier* to secure automatic undersea films.

On the first of October at 4:00 A.M., the *Élie Monnier*, glistening in a fresh application of white paint, stood out from Dakar to meet the *Scaldis*, the Belgian freighter which carried Professor Piccard, a host of scholars, and the *Bathyscaphe*.

After the rendezvous, I could wait no longer to see the *Bathyscaphe*. I lowered a launch and transferred to the *Scaldis*, where I rushed through the courtesies of Captain La Force and the scientists and scrambled down the ladder to the open cargo hold which contained the depth vehicle. Floodlights went on and I beheld the miraculous submarine.

There was a great blunted diamond shape of a metal balloon, from which depended a steel ball nearly seven feet in diameter, the yellow and white observation car in which I was to descend. On either side of the ball were electric motors, driving the propellers that would stir us around in pressure four hundred times that of the surface. I knew the principles by heart from the blueprints, and now I touched the real thing. My theoretical trust in the *Bathyscaphe* was reinforced by the machine itself. The observation sphere was molded of three-and-a-half-inch-thick steel with two heavy steel hatches framing transparent lucite cones, six inches thick.

The balloon held six steel tanks with a capacity of ten thousand gallons of extra-light gasoline with a density little more than half that of salt water. The *Bathyscaphe* did not employ gasoline as fuel, but rather as a lifting medium. Gasoline had the cardinal virtue of being a fluid with a relatively low compressibility factor; the sea could not compress petrol as it did air. The *Bathyscaphe*'s metal balloon was theoretically capable of surviving a pressure equivalent to fifty thousand feet, a depth that does not exist. We were going down to thirteen thousand feet, the average depth of the sea, leaving a safety margin.

Professor Piccard's most daring conception was one which the aqualungers of the Undersea Research Group thoroughly approved, the idea of descending without lines to the surface. He had rejected the previous design of deep-diving chambers, that of armored spheres being lowered on cables. Dr. William

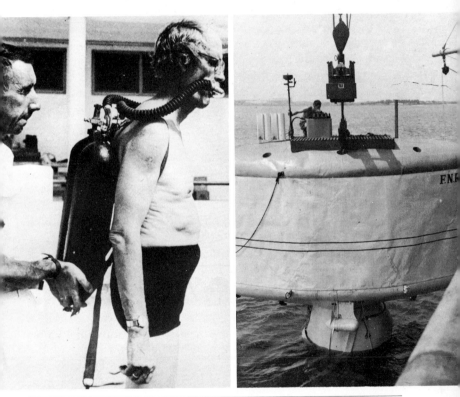

Auguste Piccard, inventor of the Bathyscaphe, the undersea dirigible, tries aqualunging at Dakar, Africa. Behind him is Tailliez. Piccard's undersea dirigible contains gasoline instead of helium. From the balloon hangs a steel observation car. It went down 4,600 feet. (Photo by Claude-Francis Beouf.)

Beebe had descended in a ball, encumbered with immense weights of cable. It had no maneuverability and each cable length increased the danger to the pilot.

The *Bathyscaphe* was to navigate twenty-five times as deep as conventional submarines. It was to go down vertically by means of iron shot ballast, which was jettisoned to retard speed on the descent, and also by valving gasoline if necessary. Under the car hung a three-hundred-pound contact weight, shaped like a skate. When it touched bottom, it decelerated and stopped the vehicle, and served as a mud shoe to permit the *Bathyscaphe* to cruise three feet from the floor at one knot in a range of ten nautical miles. Through the lucite windows the pilots could look into a landscape lighted by exterior flood-lamps, strong enough to afford color cinematography in the total blackness.

Within the depth car there was a formidable maze of controls. Our Group had built a mechanical claw with which the men inside could seize objects in the sea. There were dozens of indicators, gauges and instruments, including a Geiger counter for measuring cosmic radiations, and the most advanced oxygen generator and air purifier ever built. Two men could live inside the armored sphere for twenty-four hours.

Among the accessories of the *Bathyscaphe* was the Piccard-Cosyns depth battery we had built at Toulon. There has never been a gun like it on land. It was a sea gun. It resembled machinery for raising a small drawbridge. The battery was composed of seven 25-caliber cannon barrels, each loading a three-foot harpoon, which were fired by hydraulic pistons at the bases of the barrels. Water pressure itself built up the propulsive force as the gun went down. At a depth of three thousand feet the pilot, triggering the Piccard-Cosyns gun, could drive harpoons three inches into oaken planks fifteen feet away. On the surface the harpoons were harmless.

The depth gun was designed to secure whatever interesting animals we encountered, possibly one of the Brobdingnagian squids that haunted our imaginations. The animals could be discouraged, not only by the harpoon, but by an electrical discharge running through the harpoon line. In case the specimen resisted electrocution, the harpoon head injected strychnine. At the base of the Piccard-Cosyns gun were spring-driven reels for hauling in harpoons and monsters.

The horizontal freedom of the *Bathyscaphe* presented problems of recovering it on the surface, before its imprisoned crew ran out of oxygen. While the machine was down in the motionless element, its tenders might drift away in currents and wind, or we might lose the *Bathyscaphe* in fog. The Navy had provided for this eventuality. Special ultrasound equipment, based on underwater goniometry, was installed in the *Élie Monnier*. The frigates, *Le Verrier* and *Croix de Lorraine*, and the aircraft could track the depth balloon by a special radar mast carried atop it.

The *Bathyscaphe* returned to the surface by abandoning weights attached to it by electromagnets. Ample provision was made to send the vehicle up automatically, in case anything happened to the crew.

Our first trip was to a calm zone in the lee of Boavista Island, a volcanic peak in the Portuguese Cape Verde group. In charge were Drs. Piccard, Cosyns, Monod and Francis-Beouf; Captain La Force of the *Scaldis*, who had the responsibility of putting the *Bathyscaphe* into the sea and recovering it, and Tailliez, Dumas and I, charged with gassing and ballasting the vehicle, tracking it during the dive, getting lines on it when it returned, and handing the *Bathyscaphe* over to Captain La Force for garaging.

It soon became apparent that we would have to abandon the auxiliary equipment. There would be no time to test efficiently the mechanical claw and the depth gun. Dumas heard the fate of the depth gun with sharp disappointment. He was to be cheated of the delicious view of ten writhing hundred-foot tentacles of a giant squid, simultaneously impaled, electrocuted and poisoned two miles down.

Professor Piccard requested that the first bath of his invention be made with the inventor inside, to satisfy the interests of his sponsors and admirers. The others deferred to him. We anchored the ships in a hundred feet of water in the lee of Boavista Island and undertook the tedious details of launching the *Bathyscaphe*. Five days passed in bizarre annoyances and delays before we reached the last task, that of attaching a twelve-hundred-pound electric battery and tons of iron pigs to the vehicle by electromagnets. When the *Bathyscaphe* was scheduled to return from a deep automatic dive, all these weights were to be jettisoned by a spring-wound clock, with its "alarm" set for the appointed minute.

As the submarine hung in the hold, Professor Piccard entered the car for a last-minute check on instruments. He saw the chronometer, which was running, but there was another clock that was not. Distracted, but still a good Swiss, he wound the idle clock, without noticing that the alarm index, a red tab on the clock face, was set for twelve o'clock noon. We hung the last weights, after a morning of struggle, and were ready for the next step. At noon the weight-tripping clock went off and tons of metal fell into the hold with a frightful roar.

By a miracle, no one was under the weights. After that the *Bathyscaphe* hung in isolated majesty. The seamen had to have loud unmistakable direct orders to go near the thing. Fortunately we had another big battery to replace the one ruined in the crash.

Seven of us drew lots for the honor of accompanying Professor Piccard on the baptismal voyage. Theodore Monod drew the shortest straw. The *Bathyscaphe* entered its element at 1500 hours, November 26, 1948. Piccard and Monod were handed final comforts and good wishes and were sealed in the white ball. The laboring winch raised it into the sun and floated it in a smooth sea. We pumped gas into the tanks for three hours. During the interlude Tailliez and Dumas swam around checking on submerged equipment and exchanging gestures with the entombed savants through the thick windows. Tailliez came up and reported, "Everything okay. They're playing chess."

In due course the sun set. A launch took a tow on the *Bathyscaphe* and the *Scaldis*'s lines were cut away. The launch eased the submarine further off so it would not emerge under the hull of the mother ship. The ship's searchlight played on the *Bathyscaphe*. Quartermaster Georges stood atop the slowly submerging submarine. He looked like a man standing on the sea. Professor Piccard, imprisoned in the Atlantic, tested his submarine lights and the ocean glowed. The boatmen passed Georges more steel pellets to add to the ballasting silos. He slowly sank into the sea up to his neck, then jumped off and grabbed the boat. The *Bathyscaphe* was gone. The crews of the ships lined the rails in silence, looking at the deserted spot in the sea. The *Bathyscaphe* reappeared. Georges's own weight had made the difference in the ballast. Everyone laughed immoderately. The grinning Georges took his place in the spotlight and added enough ballast to weigh himself. The *Bathyscaphe* went under again and stayed.

Sixteen minutes later, at 2216 hours, we saw the signal tower emerge, a curious aluminum structure resembling a pagoda. We towed, pumped and hoisted for five endless hours to lower the *Bathyscaphe* to the hold and liberate the crew. Floodlights glared for the movie and still cameramen who surrounded the observation car. We unlocked the hatch and swung it open. A high leather boot came out, followed by a bare shank, another boot and leg, bathing trunks, a naked belly, and the bespectacled wild-haired pinnacle of Professor Auguste Piccard. His hand was extended, clutching a patented health drink with the label squarely presented to the cameras. Professor Piccard ceremoniously drank the product of one of his sponsors. The *Bathyscaphe* was back from the deep.

News of the accomplishment was radioed to the Belgian Government, linked with a request that additional money be allowed for an abyssal dive. The expedition sailed to Fogo Island, to await the reply, and to obtain echo-sound charts of the best areas for the big dive.

The day the *Bathyscaphe* was to make her first deep dive, without men aboard, was a Sunday, and the crew of the *Scaldis* were enjoying overtime pay. The bulky vehicle was hoisted out with all her automatic devices set, including a hanging weight which would drop all ballast when it touched the sea floor. The weight was bound with rope, so that it looked like a gigantic salami. But at a crucial moment the salami swung against a davit and dropped three tons of ballast on deck. Discouragement gripped our hearts.

It was too much for Captain La Force, who demanded that we abandon the trial before the *Bathyscaphe* crashed through the hull of his vessel. I heatedly opposed this notion. "These accidents are not due to any theoretical fault. We've got to give her another chance." The scientists, of course, backed my position. The Captain agreed to one more test. The vessels moved to Santa Clara Bay, under São Tiago Island, where the ocean was five thousand seven hundred feet deep.

At dawn Cosyns set the *Bathyscaphe*'s weight-releasing clock for eleven hours away. It was going to drop weights at 1640 hours. At 1600 hours the *Bathyscaphe* went under. The boatswain held an ax over the towline. I waved to him. He chopped the cord.

Dumas and Tailliez swam down with the *Bathyscaphe*. At one hundred and fifty feet they got the last glimpse of her rapidly disappearing into the blue. If the *Bathyscaphe* did not return, Piccard's wonderful idea was finished forever. A failure today meant that the dream of science of penetrating the last earthly unknown would be set back decades. If the *Bathyscaphe* returned, we knew that in our lifetime, depth vehicles built on her principles would carry men into the abyss.

An impressive silence ruled the ships. I pledged a bottle of cognac to the first man to spot the *Bathyscaphe*. The crew scampered up the masts and funnel, and the blue sky was dotted with the red pompons of the *matelots*. After twenty-nine minutes came an ear-splitting shout from mechanic Dudbout, "There she is!" The balloon emerged from the ocean two hundred yards off. We were so overstrained at the sight of the marvel that it took a moment to grasp a very odd fact. The well-fastened aluminum radar mast was cleanly removed, as though by a mechanic.

The divers went into the water *en masse* and raced to inspect her. I swam around the submerged machine and found her floating well with no gas leaks, but the thin plates of the balloon, especially where they passed through the surface, were rending, billowing, and sucking inward like the cheeks of an obese giant puffing on a fire.

By sundown we managed to get the *Bathyscaphe* alongside

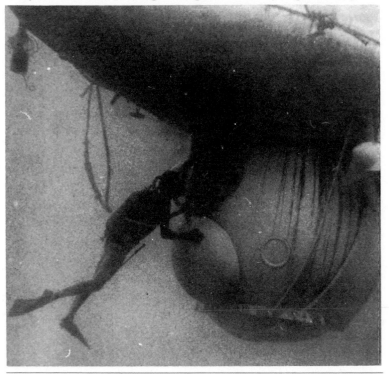

Underwater a diver checks the Bathyscaphe *before it goes down into the abyss.*

the mother ship, but the flotilla was drifting away from the shelter of the island, and we could not get a hook on the submarine. Georges and a deck mate from the *Scaldis* stood atop the balloon, trying again and again to make her fast. The submarine pitched and rolled in a freshening breeze, and we feared she would be destroyed by crashing into the *Scaldis*. Dumas and Tailliez worked all night on the *Scaldis* to avert the collision. They could not couple gas hoses to siphon her silos. Cosyns ordered the gas tanks blown with compressed carbon dioxide. Plumes of gas vapor enveloped the *Scaldis*. A spark would have touched off an explosion which would have certainly destroyed both vessels. Georges and the deck mate clung heroically to the valves, receiving jets of gasoline in their faces. They completed their work and were taken off, temporarily blind and exhausted. Through the night we fought to save the *Bathyscaphe*. She was at last lowered into her hangar in a glorious sunrise.

It sickened us to look at the vehicle of our overweening ambitions. The envelope was lacerated beyond repair. One of the motors and propellers was torn away. Inside the balloon was a mess of paint dissolved by the escaping gasoline. We opened the hatch to examine the instruments. We read the automatic gauge showing the depth attained, and made temperature and salinity corrections. The *Bathyscaphe* had reached four thousand six hundred feet.

The ironical fact was that she had survived all the pressures of the deep, with the exception of the mysterious loss of the radar mast, and then been knocked out of commission by a mild surface swell. We had the machine to carry men to the abyss, but we could not pass it through the molecular tissue of air and water.

The *Bathyscaphe* has been redesigned to make her seaworthy. She can be towed without the use of a mother ship. The

---

The Bathyscaphe *was wrecked not by pressure, but by surface waves lapping the envelope.*

pilots can enter immediately before a plunge and come out of the observation car as soon as the vessel has surfaced. There will be another trial. I am confident that the second *Bathyscaphe* will take men to the basement of the world.

Chapter Ten **Sea Companions**

O N THE *Bathyscaphe* voyage we had a few weeks to do our first work in the open Atlantic. We pored over the charts and found two magic specks labeled the Salvage Islands, between Madeira and the Canaries. The pilot book described them as uninhabited. We set course for them through the shark zone, and security precautions were in order. Divers descended in pairs to bodyguard each other, and cupric acetate shark-repellent tablets—"fly-tox" we called them—were lashed to ankles.

At the lonely island of Salvagem Grande, Didi and I went down the ladder for our first dive, he carrying a big arbalest with an explosive harpoon and I with cinécamera. We pushed off the ladder and immersed our masks. Simultaneously with desperate reflex motions, we grabbed the ladder. We had glimpsed something that aroused a fear we had never experienced before. It was pure vertigo.

We exchanged glances and cautiously put our eyes under again, holding on to the ship. We saw the bottom a hundred feet down in naked detail. There seemed no water below us. There were no motes of animals, plants or minerals in the space. It was distilled water, not the benign speck-crammed element we called clear water, in which exceptional visibility embraced an area no bigger than a concert hall. We saw a horrible bright landscape. If we let go our handholds we thought we would plummet through empty space and crash on the malign rocks that ranged far across the floor.

At last we submerged and, surprisingly, the sea supported us. We swam down, seeing each other as gross, unfamiliar animals in that surgical water. A few meters below we passed a group of motionless barracudas, which took no notice of us. Rock bass and alewives hung suspended in the spell.

The weirdest realization was the bare glossy brown whorls of the lava slope, which felt polished to the touch. Our friend, Professor Pierre Drach, had told us that there were apparently no undersea rocks and reefs on earth that were not covered with flora and fauna. Here was one exception. On the submerged slope of Salvagem Grande not one customary plant or animal appeared on the ghastly lava, save for one species that we almost ignored in disbelief. Along the cliff were immeasurable thousands of sea urchins, a large tropical variety with twelve-inch spines. We hung on our sides and stared at the clinging nation of urchins, moving their spines rhythmically like a field of wheat played by a breeze. We rolled back on our stomachs and felt the giddiness of the void. We were reassured to see our bubbles pluming toward the sky, and to find that we could actually rise and grasp the ladder. We punctured no fish and took no pictures. It was not like the sea.

On a late summer morning the *Élie Monnier* ran into Dakar in a smooth sea which we knew was a veil of illusion concealing thousands of sharks. We had been preparing for Atlantic sharks for two years. We had the finest antishark defense ever devised by the mind of man and the stout blacksmiths of Toulon. It was an iron cage, resembling a lion's cage in a sideshow, a collapsible structure which could be erected quickly and lowered into the water. It had a door through which a diver could exit and enter underwater and bar himself against sharks.

We believed that sharks were most dangerous when a diver was entering or leaving the water. Now we could be lowered safely, emerge on the bottom for our work, return to the cell, lock ourselves in and be raised in complete security. Inside the cage there was a buzzer for signaling the ship.

The grand debut of this human zoo occurred south of Madeleine Island off Dakar. Dumas, Tailliez and I, the proud designers, assumed the honor of the first dive. Encumbered heavily with three-cylinder lungs, cameras and arbalests, we entered the cage on deck and grasped the bars as the cargo boom swung us out over the water. Dangling from the end of the boom, we found the gentle roll of the *Élie Monnier* grew somewhat emphasized, like the beginning of a carnival ride. We waved good-by to our delighted shipmates and sank beneath the glassy swell.

Water is an embracing medium. It lifted us against the roof of the cell. We disengaged ourselves and floated free inside the bars. The effect was rather like a preposterous underwater birdcage, with three men tumbling in clumsy flight. The dipping ship's crane bounced the cage up and down, dealing us smart blows on the crown and feet, and swaying us against the bars violently. As the cable lengthened, the cage performed even more picturesquely. Our hollow air cylinders smote the bars, giving off loud bell-like chimes, which reverberated through the sea. Nine cylinders banging on the cage and against each other sounded like an inebriated bell ringer saluting the new year.

My mask was wrenched off and my head banged the cage. I remolded my mask, refusing to signal defeat. I buzzed to send us on to the bottom. The cable stopped with a mighty jerk and swung the battered passengers to and fro above the murky floor. We held on and gazed longingly through the bars at freedom. A bevy of brown doctor fish with bright yellow spines, all dressed up for a day at the zoo, stopped and looked in at us.

They strolled away and a six-foot barracuda appeared. It continued past without stopping, but we did not fail to appraise its circumference. The baracuda could just as easily have swum through the bars of our cell. I gave the signal to be raised.

The man cage was used only once after that, when it was lowered empty to serve as an emergency refuge. Later we met sharks without the cage.

Overleaf, We invented an antishark cage, in which Dumas goes down, feeling a little silly. The idea was to give divers a quick refuge downstairs. We abandoned the human zoo when we decided shark danger was negligible.

Our echo sounder located the wreck of a French submarine that had foundered during the war in seventy-five feet of water outside Dakar harbor. We plunged to the wreck. The vessel lay clean and upright, surrounded by such clouds of fish as we had rarely seen, small silver fingerlings and dark pomfrets. Dumas swam into the shadow of the port propeller and came face-to-face with a gigantic jewfish, a grouper variety cousin to our familiar Mediterranean merou. This specimen was ten times the size of our old acquaintances. He weighed at least four hundred pounds. The wide flat head and tiny eyes advanced on Dumas. The ugly mouth yawned open, wide enough to admit Dumas. He knew that sedentary groupers have no teeth to speak of, but it seemed that this individual might wish to swallow him unmasticated, in the fashion of the merou, which swims agape, taking in whole octopi and lobsters. Dumas had no weapon and I was out of sight, hunting camera angles.

The cavernous mouth came within two feet. Dumas sculled backward to maintain his distance and watchfulness. The monster was unhurried and Dumas kept a modest interval. Didi knew the species was harmless, but that comfort was far away as he looked into the mouth. For a long time—or so it seemed to Didi—he was pressed back while he and the grouper exchanged mutual stares of revulsion. The beast lost interest, turned aside and returned to its dim home under the lost submarine. Dumas surfaced in a reflective mood. "Imagine being swallowed by a lousy grouper," he said.

Perhaps our most beguiling companion of the sea has been the seal. Once the Mediterranean abounded in monk seals, *Monachus albiventer*, a species known in ancient times from the Black Sea to the eastern Atlantic. During the introduction of commercial sealing in the seventeenth century, the monk seals were ruthlessly exterminated by men who shared the mores of the Newfoundlander Abraham Kean, who boasted that he was the biggest animal killer in history—he had killed one million seals. But, occasionally we would hear old fishermen speak of living monk seals.

The trail of the extinct seals grew warm in La Galite, a group of tiny islands thirty-five miles north of Tunisia, which is famous for its lobsters that are kept alive in submerged pots until boats call from Tunis or Metropolitan France. La Galite had a garrulous, red-haired mayor who declared positively that he had seen living monk seals. "One evening everyone saw a seal pillaging a lobster pot near the jetty. He made such a mess that when he came up to breathe he wore the lobster pot like a queer hat." We roared at the picture. "We have all seen it," the mayor exclaimed. "I will show you the caves where the seals live."

Monsieur le Maire conducted us to three caves that had no traces of animals. At the fourth cave, Tailliez, Dumas and Marcel Ichac, the Himalayan explorer, went ashore to arouse whatever creatures inhabited it. Jean Alinat and I dived toward the cave mouth; he took a post fifteen feet in front of me behind a rock, while I trained the camera toward the underwater ramp. The shore party threw a stone into the cave. To their amazement, two big frightened monk seals lumbered out, a gray female and a huge white bull, and plowed into the water in a cascade of stones. *Monachus albiventer* had re-entered zoology.

In the shadowy cave mouth I saw a big white outline, which I took to be an unusual fish. While I was not unprepared to see a monk seal, the idea of seeing a white one was not admissible—an adult albino seal is a great rarity. Alinat was close enough to see that it was a seal and he made urgent gestures for me to start the camera. The seal stopped six feet from Alinat. The old albino was unique himself, but he had never seen a double-tailed fish that exuded clouds of bubbles. The bull stretched his paw, rolled his enormous eyes from side to side and pointed his mustache with a marvelous gesture. He swam straight for me. Alinat was close enough to caress the hoary white flank as it passed.

We hurried back aboard ship and put on dry clothes for an exploration of the cave. Inside, a flashlight inspection discovered a tunnel mouth large enough to admit a man. It was a twenty-foot crawl to the inner chamber, twenty feet in diameter and filled with a strong animal odor. In the center of the chamber our flashlights discovered the undismantled skeleton of a great monk seal. Here was the ghetto in which the species had survived, where they bore children out of sight of murdering man, and here they crawled to die when their only enemy shot them. We could not dismiss the impression that the skeleton was a tomb. It lay as well-kept as a monument.

Our search for the monk seal led us further, to Port Étienne, a French outpost near the Spanish Gold Coast. There, in a corrugated iron hut, we met a lonely man, M. Caussé, who declared that the seals were his only friends. "I have learned a whistle that brings them to me," Caussé told us. "On my Sundays, I go early and crawl quietly on the sand to their midst, and we spend the day on the beach." We looked at him and wondered which was extinct, monk seals and Caussé, or our civilization.

He was acquainted with two hundred survivors of the supposedly extinct colonies of Gold Coast seals. After he had introduced us to a herd, we put on bathing trunks to emulate our host's sociable belly crawls. Philippe and Didi, in mask and fins, swam in from the sea. They were careful about overly familiar contact with mammals twice their size, which could bite through flesh and bone with their powerful jaws. Twenty seals were bathing in the surf, including a large dark male, a mother and infant, and several playful adolescents.

Floating in the water, Dumas closely studied the seals' diving technique. They closed their nostrils, turned on their sides, caressed the water with their cheeks, and vanished without a splash. Dumas, the most "liquid" of us, looked awkward as he tried to copy them. A heavy swell labored on the rocks, shaking up muddy water full of nettling micro-organisms and stinging jellyfish, but Didi and Philippe were too absorbed in their swimming lesson to notice the inconveniences. The seals seemed to enjoy the visit of the amateurs. A big bull quietly submerged behind Tailliez and popped up to surprise him, face-to-face. Philippe cupped his hand and splashed the seal in the face. The seal puffed and blew like a small boy. Dumas flailed with laughter. The laugh turned to a shout. He rolled over and thrust his mask into the water. He saw the departing rump of a seal which had sneaked up and tickled his back with its whiskers.

I had determined when we first saw the lost colony to take a pup back to France and train him to dive with us like a hunting dog. We kidnapped an eighty-pound adolescent in a net lowered from the cliff. As we hauled the net reproachful eyes watched from the surf. Caussé's eyes expressed the same disappointment. "Don't worry," I said, "we'll take very good care of him. We'll make a friend of him."

The sailors named the pup "Dumbo." They erected the notorious antishark cage on deck and spread a rug in it for him. The pup sulked, lay prostrate and refused to eat. Dumbo had not eaten for six days when we put in at Casablanca. Worried, we tried to hire the public saltwater swimming pool to jolly the pup out of his torpor. While we were negotiating for the pool, an amiable Arab fisherman climbed aboard and looked at our dolorous seal through the bars of the former human zoo. "Say," said the visitor, "seals simply love octopus. You should try it." I clutched his arm. "Please get us some octopus."

The fisherman went ashore and cut an olive branch and lashed it to a pole. He stuck the branch into the water by the stone piers of a jetty and flirted the silvery leaves before a crevice. An octopus, thinking that the leaves were small fishes, reached out a tentacle and curled it around the lure. When the octopus had embraced it with all tentacles, our friend jerked the cephalopod out on the jetty. He took three small octopi in twenty minutes.

We threw them in Dumbo's cage. The pup bounded up and

swallowed them like spaghetti. From that moment on, Dumbo gobbled up every kind of fish we could buy. He became wonderfully gregarious, but the vigorous pup had revealed an alarming aspect of his friendship: he ate $200 worth of fish a month. We figured that he would eat a thousand dollars' worth of food a month when fully grown.

We thought of throwing him in the Mediterranean, but realized that Dumbo, now unafraid of humans, would pop up beside some fisherman, bawl for food and be slain in a panic. We couldn't ship him back to the Gold Coast. Even then, would his colony accept the sophisticated foreign traveler? We sadly decided to give him to the Marseilles zoo, where he was installed in a big pool of his own. We visited him several times. From Africa, Caussé sent him Christmas greetings. But soon Dumbo no longer recognized his benefactors of the *Élie Monnier*. He turned away from us and barked at a little old woman in black who came every day to give him a fish.

We visited the luxuriant waters of the Cape Verde Islands where every dive was a marvel for us. We imagined what observations Charles Darwin, who explored there in 1831 during the famous oceanographic voyage of H.M.S. *Beagle*, could have made with our equipment. "While looking for marine animals, with my head about two feet above the rocky shore," Darwin wrote, "I was more than once saluted by a jet of water, accompanied by a slight grating noise. . . . I found out that this was a cuttle fish. . . . I observed that one which I kept in the cabin was slightly phosphorescent in the dark."

We were permitted to observe the cuttlefish, or octopus, with our heads below the water. We saw the great rays and mantas swimming in the depths. Under Boavista Island, blue

On a remote beach of the Spanish Gold Coast, West Africa, we find a colony of monk seals, supposed to have been exterminated in the 1690s. Dumas and Tailliez crawl up on the beach like fellow seals and get acquainted. Below, Dumas and Tailliez go into the surf with the "extinct" seals and play follow-the-leader. The seals would duck under and tickle the divers with their whiskers.

lobsters were so numerous that there were not enough crevices to house them. The homeless ones wandered on the bottom along crowded boulevards between the dwellings of the home owners. It looked like slow-motion city traffic.

The turtles of Brava astonished us by the duration of their submersions. A captured turtle in a zoo pool surfaces often for breath, but here in the wild they rested on the bottom for hours. Only once did we see a turtle ascend to breathe. Their metabolism is possibly so low that turtles require little oxygen, except when paddling energetically to escape.

Fifty feet down off Brava we discovered a large tunnel which passed completely through a small island. In the dark interior, one could look back reassuringly at the emerald glow of the entrance, swim on through shafts of silvery light falling from potholes in the rock above, and turn a corner to find the inviting green of the sea at the far exit. The cave entrances were exuberant with bright silver-blue fish, as animated as a wedding party. The gatherings were exactly that; a mass marriage of large blue jacks was taking place. Their bellies were dilated with eggs. Jacks were everywhere in the waters outside, in swiftly moving forage parties of four to thirty fish, but here they gathered by the hundreds in a glittering mass circling in the shadows. They were quite excited by the invasion of the divers, and gathered around us reprovingly, like guests at a genteel reception glaring at uninvited drunks.

We would enter the cave slowly from dark corners so as not to disturb them, and witness the excited flutterings of their long love dances in the nuptial suite. We made ourselves as inconspicuous as possible out of respect for one of nature's secret ceremonies, perhaps never before witnessed by man.

Our best companion in the reefs was the tragicomic trumpet fish, found everywhere in great numbers in the Cape Verdes. The trumpet fish has a head like a horse, and a disproportion-

Above, "Dumbo," a youngster we took back to the Marseilles zoo. Below, this wary Cape Verde surgeon fish gets his name from the razor-sharp retractable scalpel he carries at the base of his tail fin.

ately tiny tail, separated by a long pipe of a body, sometimes two feet in length. The hapless trumpet fish, more properly called *poisson flute*, or flute fish, is wretchedly equipped for locomotion. The useless tail and the stiff tubular body are handicaps for which the trumpet compensates by wildly agitating its pectoral fins to move forward and backward, slanted head down, or leaning on end as frequently as it swims horizontally. From a rocky hole a dozen of the poor chaps will be found sticking up like pencils in a cup.

These forlorn sticks have a remarkable trait. We observed it at such length that what I report here is not a hasty conclusion but a confirmed observation, recorded many times by the cinécamera.

Often a trumpet fish will leave his fellow wallflowers and swim rapidly toward a larger animal such as a parrot fish, a grunt, grouper or rock bass. He will place himself alongside or on the back of the passer-by, barely touching it, and swim along in this tandem fashion as if seeking friendship, begging for tenderness, offering his heart. There is absolutely no hostility in this gesture. The trumpet fish has no weapons to harm a fish his size, and is in fact placing himself in some danger from the other more capable fish. Nor is the trumpet trying to beg part of the other fellow's dinner.

The gesture is never reciprocated. The rock bass or parrot fish goes about his affairs, heedless of the trumpet fish, until he grows annoyed at the bore pressing his unwanted friendship. He makes a sudden spurt to dislodge the nuisance, but the lonely one follows. Later, the object of affection accelerates to full speed and leaves the trumpet fish hanging morosely in the water, rejected once again. We watched this piscine social drama many times, torn between pity and laughter.

Above, in the Cape Verdes we met the thin lonely trumpet fish and almost lost our mouthpieces laughing at his attempts to make friends with the coral-eating parrot fish. Below, Dumas tickles the belly of a Pei-qua, the Mediterranean rockfish.

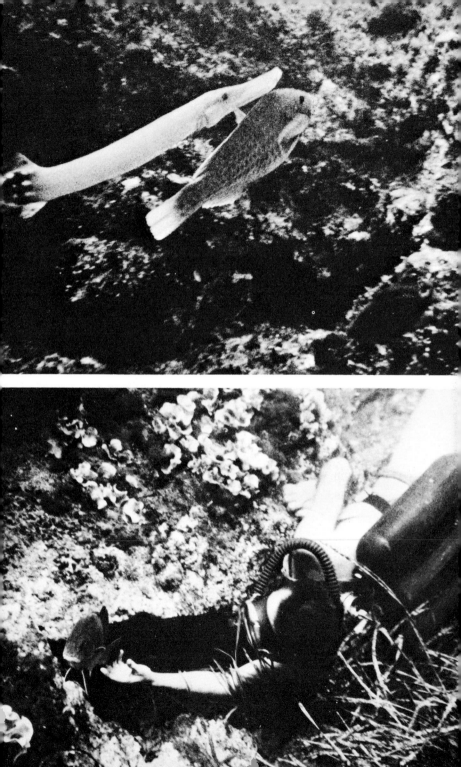

The Gibraltar Straits are a unique place to study sea mammals. Migratory thousands of whales and porpoises pass to and fro across the narrow sill between the Mediterranean and Atlantic. Tailliez and I watched the herds romping through the gate, while Dumas was under the keel fitting an automatic movie camera to film an antic porpoise pack playing at the bow of the *Élie Monnier*.

One watches them race the bow of a ship, vaulting out of the foam for breath and dropping out of the stream like a man falling from flight, to be replaced instantly by another porpoise; and, as the bow speeds, lie on their flanks and spy the humans with quick little eyes. A mother swims with her child, which moves at a faster rhythm to keep up; they jostle each other playfully. Presto, for no apparent reason, the ranks thin out — the last porpoise sounds and a curtain of foam is drawn over the ballet of the sea.

We often watched them and occasionally dived with them. They played chasing games as if they had a brain capacity for satire. They are constructed disturbingly like men. They are warm-blooded and breathe air and are the size and weight of men. Dr. Longet dissected a porpoise on an operating table on deck. We watched uneasily as he removed lungs like ours and a brain as big as a man's, deeply corrugated in the fashion that is supposed to mark human genius. Porpoises have smiling lips and shining eyes. They are gregarious and, more than that, social. There are probably more porpoises in the sea than there are men on earth.

The powerful horizontal flukes of the porpoises speed them to the surface to take an instant breath, then they dive like a living torpedo. We took slow-motion movies of their ventholes to measure the time they need to snatch a breath. The films showed that they fill their lungs in one-eighth of a second. When they went down they left a silvery dotted line of bubbles, revealing that porpoises do not hermetically seal their blowholes under water.

Swimming under water among them with naked ears we

heard their mouselike squeaks, a comical cry for such splendid animals. The porpoise's shrill pip may have a further use than mere conversation in the herd. One day, forty miles out in the Atlantic on a course for Gibraltar, the *Élie Monnier* was running its honest twelve knots when a crowd of porpoises overtook from astern. They were headed on the exact bearing for the center of Gibraltar Strait, although land was far out of sight. I ran with them for a while, then subtly altered course five or six degrees, trying to deviate them. The pack accepted the detour for a few minutes, then abandoned the bow and resumed its original heading. I swung back on the porpoise course—they were running true for the Straits.

Wherever they came from the porpoises had secure knowledge of where the ten-mile gate lay in the immense sea. Are the porpoises equipped with sonic or ultrasound apparatus by which their squeaks give them the feel of unseen bottom topography? Things happen exactly as though they receive bearings from Gibraltar. They probably bounce their mouselike chirps from the floor. Perhaps deep in their racial instinct there is a knowledge of the course which winds through the far unseen hills and plains to the door of their Mediterranean playland.

Chapter Eleven **Monsters We Have Met**

FISHING is one of man's oldest occupations and fish stories entered folklore very early. Poets and nature fakers added their touches to marine superstitions that persist to our day. The popular press still cannot resist unsubstantiated stories of sea monsters.

When the helmet diver appeared a century ago, the saga gained the ultimate dramatic ingredient, a human hero to descend and give battle to the fiends. Their sanguinary engagements have been portrayed by dry writers ashore. The lonely, hard-working divers may be forgiven for their silent endorsement of the sagas. Indeed the helmet diver, imprisoned in his casque, and almost always working in filthy harbors and channels, is unable to determine whether an interference with his air pipe is caused by a giant squid or a rotted spar. Doubt leaveroom for interpretation.

A naked man swimming in the sea mingles with and observes life around him and may be watched by other swimmers, and the recording eye of the lens. His advent means the end of superstition.

If I may put aside the sea snake, the villains of undersea myth are sharks, octopi, congers, morays, sting rays, mantas, squids and barracudas. We have met all but the giant squid, which lives beyond our depth range. Save for the shark, about which we are still puzzled, the monsters we have met seem a thoroughly harmless lot. Some are indifferent to men; others are curious about us. Most of them are frightened when we approach closely. I write here of some of our "monsters"—and of the shark later.

Our experiences, of course, have been mainly in the Mediterranean with shorter periods in the Atlantic and Red Sea. Perhaps the monsters of the Mediterranean have been tamed, and the wild ones live in your ocean. Consider the case of the slandered octopus.

The octopus owes most of its notoriety to Victor Hugo, who, in *Toilers of the Sea*, related the manner in which the octopus ingests food, in this case a human being. "You enter in the beast," he wrote. "The hydra incorporates itself with the man; the man is amalgamated with the hydra. You become one. The tiger can only devour you; the devilfish inhales you. He draws you to him, into him; and, bound and helpless, you feel yourself emptied into this frightful sac, which is a monster. To be eaten alive is more than terrible; but to be drunk alive is inexpressible." Such was the anticipation of the octopus we took to our first dives. After meeting a few octopi, we concluded that it was more likely that to be "drunk alive" referred to the condition of the novelist when he penned the passage, than to the situation of a human meeting an octopus.

On countless occasions we have offered our persons for this libation. At first we had natural revulsion against touching the slimy surfaces of rocks and animals, but found that the fingertips conveyed no such sense. That made it easier to touch a live octopus for the first time. We saw many octopi on the floor and clinging to reefs. Dumas seized the nettle one day, by pulling an octopus from a cliff. He was somewhat apprehensive, but it was a small octopus and Didi felt he was too large a drink for it. If Dumas was timid, the octopus was downright

---

"No fate could be more horrible than to be entwined in the embrace of those eight clammy, corpselike arms, and to feel their folds creeping and gliding around you, and the eight hundred discs with their cold adhesive touch, gluing themselves to you with a grasp that nothing could relax, and feeling like so many mouths devouring you at the same time." —Victor Hugo

Below, Guy Morandière pulls an octopus off a reef 90 feet down and tries to induce the shy animal to play with him. The octopus has never read Victor Hugo.

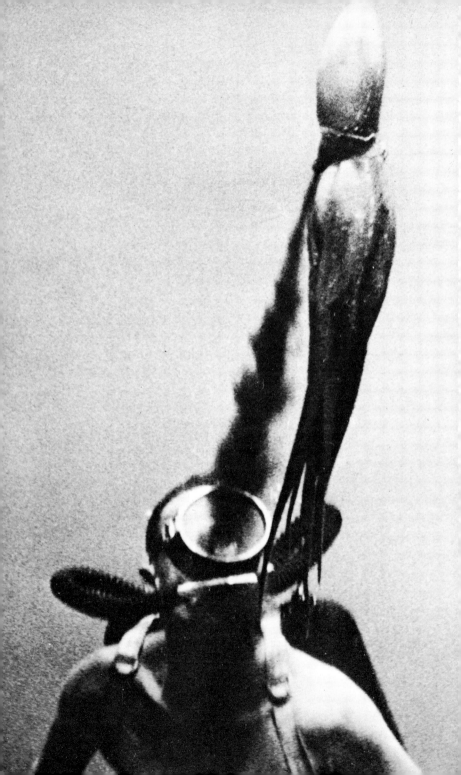

terrorized. It writhed desperately to escape the four-armed monster, and succeeded in breaking loose. It made off by slow jet propulsion, exuding spurts of its famous ink.

Soon we were handling any size of cephalopod we found. Dumas became a sort of dancing instructor to devilfish. He would select an unwilling pupil, hold it firmly and gently and gyrate around, inducing the creature to follow. The octopod used every trick to escape. The bashful animal usually refused to fasten its suction cups to flesh. Didi tried to wrap the tentacles around his bare arm, in the familiar blood-drinking position, but without success. The octopus would not retain the grip. Didi forced the suction cups against his arm and succeeded in obtaining a brief adhesion, quite easy to remove, leaving momentary marks on his skin.

The octopus has a remarkable trace of adaptability. Dumas determined that, by patiently playing with them until he met some response. Usually octopi were most submissive when very tired. Dumas would release an exhausted octopus and let it jet away with its legs trailing. The octopus has two distinct means of locomotion. It can crawl efficiently on hard surfaces. (Guy Gilpatric once saw an octopus let loose in a library. It raced up and down the stacks, hurling books on the floor, possibly a belated revenge on authors.) Its method of swimming consists of inflating the head, or valva, with water and jetting the fluid to achieve moderate speed. Dumas could easily overtake the animal. The octopus discharged several ink bombs and then resorted to its last defense, a sudden plunge to immobility on the bottom, where it instantly assumed the local color and pattern. Keeping a sharp eye out for this camouflage stunt, Didi confronted the creature again. At the exhaustion of its psychological warfare effects, the octopus sprang hopelessly from the bottom, fanned its legs and dribbled back to the floor.

At this point Dumas found it willing to dance. Taking the

A bashful octopus refuses to dance with Dumas and is jet-propelling himself away in a cloud of ink.

student by the feet, he led it through some ballet improvisations. Several octopi induced to this state of nervous collapse responded imitatively to his figures, and ended the lesson in the attitude of a playful cat. When Didi's air was gone, the spent octopus remained extended and relaxed, watching him fly into the sky. I know this sounds like a story from Marseilles. I was careful to make several movies of it as evidence.

The ink of the octopus has been liberally diluted with journalistic fantasy. Masks protect our eyes so I cannot say whether or not the ink is optically venomous. It had no effect on naked skin and appeared to have none on a fish passing through the ink. We found that the emission was not a smoke screen to hide the creature from pursuers. The pigment did not dissipate; it hung in the water as a fairly firm blob with a tail, too small to conceal the octopus. If the ink wasn't poison or concealment, what was its function? I heard an interesting explanation from a staunch friend of the octopus, Theodore Rousseau, curator of painting at the Metropolitan Museum of Art in New York. He submitted that the ink bomb is a mock-octopus shape to divert weak-eyed pursuers. The size and shape of the puff roughly corresponds to that of the swimming octopus which deposited it.

On the flat shallow floor northeast of Porquerolles we came upon an octopus city. We could hardly believe our eyes. Scientific credence, confirmed by our own experiences, holds that the octopus lives in crannies of rock and reef. Yet here were strange villas, indisputably erected by the octopi themselves. A typical home was one roofed with a flat stone two feet long and weighing perhaps twenty pounds. One side of the stone had been raised eight inches and propped by two lintels, a stone and a red building brick. The mud floor inside had been excavated five inches. In front of the lean-to was a wall of accumulated debris: crab and oyster shells, stones, shards of man-made pottery, sea anemones and urchins. A tentacle extended from the dwelling and curled around the rubble, and the owl-like eyes of the octopus peered at me over the wall. When I went closer, the tentacle contracted, sweeping the debris up against the door, concealing the inhabitant. We made color photographs of an octopus house.

To me the observation was noteworthy, for it may prove that the octopus is capable of using tools, which involves complex conditioned reflexes which I have not seen previously credited to the octopus. By assembling materials to build a house and by lifting the rock and holding it while it inserted the pebble and brick pillars, the octopus may have promoted itself in the brain classification of species.

It is intriguing to speculate on octopus love-making, which we have never seen in the sea. It was described by Henry Lee, late keeper of the Brighton (England) Aquarium, the Boswell of the octopus. Eighty years ago Henry Lee patiently observed the first captive English octopus in a tank at Brighton. He wrote a profoundly witty book called *The Octopus, the Devilfish of Fact and Fiction*. For his Victorian audience Lee wrote, "I can say but little concerning the fertilization of eggs of the octopodae in a book intended for readers of all classes." With this pious disclaimer, Henry Lee then proceeded to describe what he had seen: "In the breeding season a curious alteration takes place in one of the arms of the male octopus. The limb becomes swollen, and from it is developed a long wormlike process, furnished with two longitudinal rows of suckers, from the extremity of which extends a slender elongated filament. When its owner offers his hand in marriage to a lady octopus she accepts, *and keeps it, and walks away with it*, for this singular outgrowth is then detached from the arm of her suitor, and becomes a moving creature, having separate life, and continuing to exist for some time after being transferred to her keeping."

A favorite haunt for another breed of monster was an encampment one hundred and thirty feet down in La Sèche du Sarranier in the Côte d'Azur. The soil was distinctive—it seemed sandy until we drew near and saw it was a field of queer round pebbles of organic origin, tinted in delicate shades of rose and mauve. There were a few stone cairns, inhabited by merous and rockfish, but the place was owned by rays. A host of sting rays, eagle rays and skates rested flat on the pebbles.

As we swam to them, they raised alertly on their wingtips, ready to flee, and when we closed in, they rose in pairs and fled. We often saw them swimming in couples, but we have not been able to capture a natural pair to see if they were sexual mates. Once I came upon two medium-sized sting rays asleep on the bottom. One awoke and started to fly. It hesitated, returned to the other and awakened it by flapping its wings. They sailed away together.

When we glided motionless into the ray kingdom, they remained, rolling their big round eyes and closely watching us. The thicker bodies were pregnant females, which retain their young for a long time, as if to launch them as capable as possible in the struggle for life. Spearing rays has no further interest for us. The killing is simple and unworthy. In the early days we sometimes harpooned rays. One that we landed surprised us by giving birth on the sand. Tailliez picked up one of the eight-inch calves to return it to the water. The newborn infant gave him a man-sized sting.

Fishermen are sometimes hurt by boated rays and observe the rule of severing the tail when the animal is hauled in. The wounds are often infected. There is a poison gland in the tail, and the thick coat of mucus on the serrated stinger may plant infection in a wound.

Rays are no danger to a diver. Certainly the ray will never attack a man. The celebrated stinger is not an offensive weapon —it is a reactor to molestation. The stinger is located at the base of the tail, extending for only a sixth of its length. Dumas swims up behind a ray and grabs the end of the tail, an insurance against an accidental sting. The ray struggles to release its tail from his grasp, but it cannot manipulate the stinger while Dumas holds the tip. The saw-toothed weapon is placed to defend attack from behind and above. Bathers who step on a ray may receive this reflex stroke, inflicted as deeply as the frightened animal can swing. It may mean several weeks in the hospital.

While we were diving off Praia in the Cape Verdes, a shadow passed across the bottom. I thought it was a cloud scudding in the other world, until Dumas hooted and pointed up. Directly over us passed a manta ray with an eighteen-foot wingspread. It eclipsed the sun. It did not swim—it flew. The curved extremities of its wings sliced the surface film. The belly was enamel white, as white as the back was black. The supernatural sight did not remain long. Gliding without apparent effort the manta drew away from Dumas's best two-knot pursuit, flickered its wings and accelerated into the sea fog.

Fishermen fear manta rays, a supersition which is enhanced by its nocturnal prank of a violent leap into the air and a flat resounding crash of a ton of flesh on the water. Fishermen warned us that mantas killed divers by wrapping their wings around the man and smothering him, or by enveloping the diver and crushing him against the floor. But far from inspiring fear in a diver, they arouse admiration in a man lucky enough to see them in flight. We dissected a manta ray to examine its nutritive system. It had no teeth or grinders. Food was taken by a powerful water pump comprising the gill clefts and mouth, drawing large volumes of water into an elaborate filtering system, which strained morsels of plankton as fine as pablum to pass them through the little throat. Unlike the sting ray it had no stinger on its tail. The manta ray had to rely entirely on its speed to survive. It could harm nothing but plankton.

At Brava Island Dumas penetrated the camouflage of a large sea tortoise which clung to a rock, secure in its blend with the reef. He approached from the rear and grasped the shell by either rim. The astonished turtle jerked its flippers. Didi lifted

---

Overleaf, this big manta ray looked to me like a twin-engined bomber as we met under the Cape Verdes in 1948. Following pages, Dumas goes after a sting ray 120 feet down off Porquerolles, France. He grabs them in the safest place, the tail tip. The stinger is near the root of the tail. If you grab the tip, the ray can't sting you.

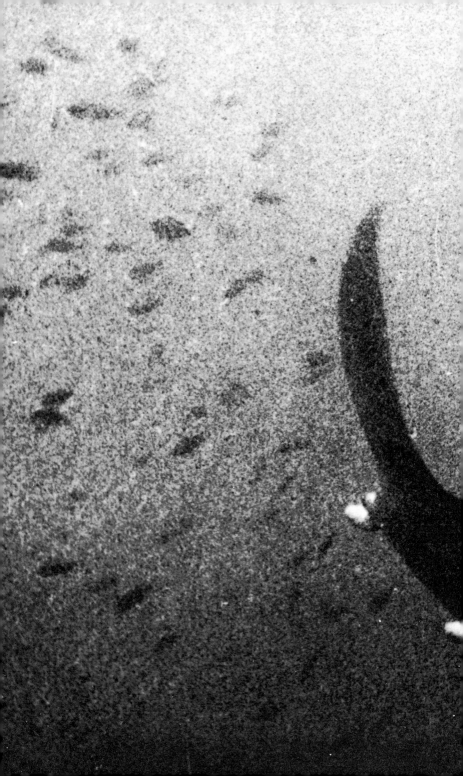

the animal and applied a slight propulsion with his fins. The outraged turtle took up the tandem glide and paddled Didi around a vertical loop. Dumas experimented with several aeronautical figures, including a worthy Immelmann, and liberated his towboat. The turtle did not understand freedom. It repeated the last loop like a comedian performing a double take, before it rowed off into the green.

In undersea fiction, the moray eel is a formidable gangster of the deep. It guards as many sunken treasures as does the literary octopus. Fishermen fear the moray on a realistic basis. Flopping out its life in the bilge boards of a boat, the moray *in extremis* will bite anything presented to its jaws. Wise fishermen crush its head as soon as it is boated. Roman historians relate that Nero threw slaves into pools of morays to amuse his friends with the sight of human beings eaten alive. This celebrated perversity, whether true or not, gave the moray a bad name for all time. Nero must have methodically starved captive morays until the fish had no choice of menu.

Morays will not attack men in the sea. They presented themselves to us with only the head and neck emerging from the hole. They looked quite fearsome. In addition to speed, camouflage and weapons, fish employ psychological effects. The moray disseminates propaganda with its evil eyes and bared fangs. If it could hiss like a wildcat, it would. The moray is also found in sunken ships, staring with basilisk eyes from encrusted aeries of pipes, and trunks. Alas, it is as prosaic as you and I and the cat. It wishes to be unmolested in the destined journey of life. It is a confirmed home body. It will hence inflict a bite on an intruder. Dumas was once reaching into the reef for lobsters under Machado light, when he took a moray bite

Above, I met this sea turtle pawing a reef off West Africa. Turtles have never tried to bite us, even when Dumas grabs one by the shell to take a fast ride. Below, down in a Cape Verde reef, we found lots of angry-looking moray eels, like this one at the door of his 15-fathom home. When we grope into crevices for lobsters and happen to disturb a moray, we could get bitten.

on his finger. The puncture was unimportant and healed over-night. The next day the wound hemorrhaged and closed again. Dumas said, "The moray did not attack me. It warned my hand to get out and stay out." There was no infection. The bite was not venomous.

While we were grubbing in the harbor of ancient Carthage, we called on Dr. Heldt, director of the oceanographic station at Salambo. He and his wife had great enthusiasm for Tunisian marine fauna, and urged us to visit one of the most horrible and grand sights we would ever see, the Madrague of Sidi Daoud. A *madrague* is a gigantic tuna net originated centuries ago in the Aegean and Adriatic and brought later to Tunisia. It is a wide-meshed vertical net a mile or two in length which is stretched diagonally from the shore, terminating at sea in four roomy chambers in which big tuna are trapped during the early summer spawning season.

Tunas are migrants; some zoologists believe they cruise around the world. World travelers, or citizens of a single ocean, tuna invariably come inshore in the spawning season and swim in schools along the beaches. They always navigate with their right eyes toward the shore. Aristotle, who was no mean oceanographer, concluded that tuna were blind in the left eye, a belief still prevalent among contemporary Mediterranean fishermen. But whatever the reason honeymooning tunas keep the shore to starboard, it is the characteristic that signs their doom.

When the herd encounters the *madrague*, it turns left along the net wall to skirt the obstacle, and passes straight into the trap. Arab fishermen in boats watch the foyer of the trap and close the door when the fish have entered. They admit the tuna to a second room and close it, so that the outside door may be open for new arrivals. The fish are conducted into a third cham-ber, beyond which hangs the death cell itself, which is termed by an ominous Sicilian word, the *corpo*. Sixty gigantic tuna and hundreds of bonitos were in the *corpo* when we arrived in the village of Sidi Daoud to film the massacre in color.

The *corpo* had been towed inshore. On the jetty stood the master of ceremonies and head executioner, the *raïs*, a majestic individual in a red fez and American Army trousers. He hoisted a flag as the signal for the *matanza* (massacre). Hundreds of Arabs converged in steady flat-bottomed rowboats and disposed them in a hollow square around the *corpo*. The *raïs* was rowed to the center. He ordered the ritual to begin. A barbarian roar broke from the fishermen and they chanted an old Sicilian song, traditional to the *matanza*. To its cadence the boatmen hauled in the walls of the net.

Marcel Ichac filmed the spectacle from a boat above the *corpo*, while Dumas and I dived into the net to record it below. Sunk in the crystalline water we could not see both sidewalls of the *corpo*, and imagined that the fish could not, either. We had unconsciously taken on the psyche of the doomed animals. In the frosty green space we saw the herd only occasionally. The noble fish, weighing up to four hundred pounds apiece, swam around and around counter-clockwise, according to their habit. In contrast to their might, the net wall looked like a spider web that would rend before their charge, but they did not challenge it. Above the surface, the Arabs were shrinking the walls of the *corpo*, and the rising floor came into view.

Life took on a new perspective, when considered from the viewpoint of the creatures imprisoned in the *corpo*. We pondered how it would feel to be trapped with the other animals and have to live their tragedy. Dumas and I were the only ones in the creeping, constricting prison who knew the outcome, and we were destined to escape. Perhaps we were oversentimental but we were ashamed of the knowledge. I had an impulse to take my belt knife and cut a hole for a mass break to freedom.

The death chamber was reduced to a third of its size. The atmosphere grew excited, frantic. The herd swam restlessly faster, but still in formation. Their eyes passed us with almost human expressions of fright.

My final dive came just before the boatmen tied off the *corpo* to begin the killing. Never have I beheld a sight like the death cell in the last moments. In a space comparable to a large living room tunas and bonitos drove madly in all directions. The tuna's right-eyed honeymoon instinct was at last destroyed. The fish were out of control.

It took all my will power to stay down and hold the camera into the maddened shuttle of fish. With the seeming momentum of locomotives, the tuna drove at me, head-on, obliquely and crosswise. It was out of the question for me to dodge them. Frightened out of sense of time, I heard the reel run out and surfaced amidst the thrashing bodies. There was not a mark on my body. Even while running amok the giant fish had avoided me by inches, merely massaging me with backwash when they sped past.

The nets were pegged, the *raïs* gave the ceremonial sign of execution. He lifted his fez and saluted those who were about to die. The fishermen struck at the surfaced swarm with large gaffs. The sea turned red. It took five or six men whacking gaffs into a single tuna to draw it out, flapping and bending like a gross mechanical toy. The boats rocked with convulsive bleeding mounds of tuna and bonitos. The fish ended their struggles, and the bloody fishermen leaped into the pink water of the *corpo* to wash and relax.

More than any other fish, the great liche is the ocean's nobility, aloof from nets and hooks, living in liberty below fifteen fathoms, deigning to touch the terrestrial world at remote capes, lonely reefs and deep wrecks. The liche is the color of the ocean; elongated, powerful, swift and lissome, with a lemony stripe on its seasilver sides. Sometimes alone, more often in troops, the liche appears from liquid nothingness into the diver's world and casts its fawn's eye upon him. Then all other fish seem shabby provincials. The liche is a cosmopolite, unimpressed on its long passage from Sidon to the Pillars of Hercules by the sight of men, who are one day worth an investigatory pause and the next are pedestrians to be shouldered out of the way.

We have seen them counterlit against the sky, circling a rock needle aureoled in foam. Like tunas, liches are big, migratory carnivores. Men betray the tuna with nets and catch them on lines, but the liche will not be taken. So successfully do they elude hooks and trawls that fishermen and scientists believe the liche is rare and does not exceed three feet in length. Yet, we have counted far more liche than tuna, and are not surprised to see six-footers. The sight of liche is to us the badge of an adventurous dive.

Barracudas are no danger to divers. Despite undersea fairy tales, I know of no reliable evidence of a barracuda's attacking a diver. Many good-sized barracudas passed us in the Red Sea, in the Mediterranean and the tropical Atlantic, giving no sign of aggressiveness.

A diver is too busy avoiding a certain truly dangerous undersea animal to fret over barracudas. This real-life peril of the deep is the commonplace sea urchin, a burrowing thistle-like echinoderm with sharp, brittle spines. It is in no way aggressive, it is merely omnipresent. The urchin may not measure up to the demands of the monster-mongers, but when one bumps into an urchin there is villain enough. Its spines penetrate the flesh and break off. They are extremely difficult to remove and may become infected. We keep a sharper eye out for sea urchins than we do for barracudas.

A larger nuisance is the stinging jellyfish, whose varicolored crystal cups hang in the water like small naval mines. They are pleasingly patterned in dark blue, brown and yellow. Many varieties of jellyfish can deal a shocking sting. The most prevalent and dangerous is the Portuguese man-of-war, whose arrival at the seashore has spoiled many a resort season. The animal floats on the surface dangling its long poisonous filaments. I made a dive off Bermuda, through a colony of men-of-war, so crowded together it was hard to find a place to enter. Safe

As we cross the coral hills we stay clear of a menace worse than sharks, the reaching fingers of fire coral. Don't touch corail de feu *if you wish to avoid weeks of pain. (Photo by Dr. Nivelleau de la Brunière.)*

below the surface, I looked up at a ceiling of injurious threads, fringing the sky to the limit of sight. Among the filaments swam small Nomeus fish, of the perch-pike family, who have an absolution from the man-of-war. It never stings them.

Two important living enemies of undersea man are fire coral and sea poison ivy, which inflict burns that may last for days. They are allergenic phenomena—a few persons are immune and others suffer no pain on the first contact, but the second exposure brings a severe rash. Anti-histamine creams heal the burns of sea ivy and fire coral in a few hours.

Such are some of the monsters we have met. If none have eaten us, it is perhaps because they have never read the instructions so generously provided in marine demonology.

Chapter Twelve Shark Close-ups

O N A GOGGLE dive at Djerba Island off Tunisia in 1939 I met sharks underwater for the first time. They were magnificent gun-metal creatures, eight feet long, that swam in pairs behind their servant remoras. I was uneasy with fear, but I calmed somewhat when I saw the reaction of my diving companion, Simone. She was scared. The sharks passed on haughtily.

The Djerba sharks were entered in a shark casebook I kept religiously until we went to the Red Sea in 1951, where sharks appeared in such numbers that my census lost value. From the data, covering over a hundred shark encounters with many varieties, I can offer two conclusions: The better acquainted we become with sharks, the less we know them, and one can never tell what a shark is going to do.

Man is separated from the shark by an abyss of time. The fish still lives in the late Mesozoic, when the rocks were made: it has changed but little in perhaps three hundred million years. Across the gulf of ages, which evolved other marine creatures, the relentless, indestructible shark has come without need of evolution, the oldest killer, armed for the fray of existence in the beginning.

One sunny day in the open sea between the islands of Boavista and Maio, in the Cape Verde group, a long Atlantic swell beat on an exposed reef and sent walls of flume high into the air. Such a sight is the dread of hydrographers, who mark it off sternly to warn the mariner. But the *Élie Monnier* was attracted to such spots. We anchored by the dangerous reef to dive from the steeply rolling deck into the wild sea. Where there is a reef, there is abundant life.

Small sharks came when we dropped anchor. The crew broke out tuna hooks and took ten of them in as many minutes. When we went overside for a camera dive, there were only two sharks left in the water. Under the racing swell we watched them strike the hooks and thrash their way through the surface. Down in the reef we found the savage population of the open ocean, including some extremely large nurse sharks, a class that is not supposed to be harmful to man. We saw three sharks sleeping in rocky caverns. The camera demanded lively sharks. Dumas and Tailliez swam into the caves and pulled their tails to wake them. The sharks came out and vanished into the blue, playing their bit parts competently.

We saw a fifteen-foot nurse shark. I summoned Didi and conveyed to him in sign language that he would be permitted to relax our neutrality toward sharks and take a crack at this one with his super-harpoon gun. It had a six-foot spear with an explosive head and three hundred pounds of traction in its elastic bands. Dumas fired straight down at a distance of twelve feet. The four-pound harpoon struck the shark's head and, two seconds later, the harpoon tip exploded. We were severely shaken. There was some pain involved.

The shark continued to swim away, imperturbably, with the spear sticking from its head like a flagstaff. After a few strokes the harpoon shaft fell to the bottom and the shark moved on. We swam after it as fast as we could to see what would happen. The shark showed every sign of normal movement, accelerated gradually and vanished. The only conclusion we could draw was that the harpoon went clear through the head and exploded externally, because no internal organ could survive a blast that nearly incapacitated us two harpoon lengths away. Even so, taking such a burst a few inches from the head demonstrated the extraordinary vitality of sharks.

One day we were finishing a movie sequence on trigger fish when Dumas and I were galvanized with ice-cold terror. It is a reaction unpleasant enough on land, and very lonely in the water. What we saw made us feel that naked men really do not belong under the sea.

At a distance of forty feet there appeared from the gray haze the lead-white bulk of a twenty-five-foot *Carcharodon carcharias*, the only shark species that all specialists agree is a confirmed maneater. Dumas, my bodyguard, closed in beside me. The brute was swimming lazily. In that moment I thought that at least he would have a bellyache on our three-cylinder lungs.

Then, the shark saw us. His reaction was the last conceivable one. In pure fright, the monster voided a cloud of excrement and departed at an incredible speed.

Dumas and I looked at each other and burst into nervous laughter. The self-confidence we gained that day led us to a foolish negligence. We abandoned the bodyguard system and all measures of safety. Further meetings with sharp-nosed sharks, tiger sharks, mackerel sharks, and ground sharks, inflated our sense of shark mastery. They all ran from us. After several weeks in the Cape Verdes, we were ready to state flatly that all sharks were cowards. They were so pusillanimous they wouldn't hold still to be filmed.

One day I was on the bridge, watching the little spark jiggle up and down on the echo-sound tape, sketching the profile of the sea floor nine thousand feet below the open Atlantic off Africa. There was the usual faint signal of the deep scattering layer twelve hundred feet down. The deep scattering layer is an astounding new problem of oceanography, a mystifying physical mezzanine hovering above the bedrock of the sea. It is recorded at two to three hundred fathoms in the daytime and it ascends toward the surface at night.

The phenomenon rises and falls with the cycle of sun and dark, leading some scientists to believe it is a dense blanket of living organisms, so vast as to tilt the imagination. As I watched the enigmatic scrawls, the stylus began to enter three distinct spurs on the tape, three separate scattering layers, one above the other. I was lost in whirling ideas, watching the spark etch the lowest and heaviest layer, when I heard shouts from the deck, "Whales!" A herd of sluggish bottlenosed whales surrounded the *Élie Monnier*.

In the clear water we studied the big dark forms. Their heads were round and glossy with bulbous foreheads, the "bottle" which gives them their name. When a whale broke the surface, it spouted and the rest of the body followed softly, stretching in relaxation. The whale's lips were curved in a fixed smile with tiny eyes close to the tucks of the lips, a roguish visage for such a formidable creature. Dumas skinned down to the harpoon platform under the bow while I stuck a film magazine in the underwater camera. The whales were back from a dive. One emerged twelve feet from Dumas. He threw the harpoon with all his might. The shaft struck near the pectoral fin and blood started. The animal sounded in an easy rhythm and we paid out a hundred yards of harpoon line, tied to a heavy gray buoy. The buoy was swept away in the water—the whale was well hooked. The other whales lay unperturbed around the *Élie Monnier*.

We saw Dumas's harpoon sticking out of the water; then it, the whale and buoy disappeared. Dumas climbed the mast with binoculars. I kept the ship among the whales, thinking they would not abandon a wounded comrade. Time passed.

Libera, the keen-eyed radio man, spotted the buoy and there was the whale, seemingly unhurt, with the harpoon protruding like a toothpick. Dumas hit the whale twice with dum-dum bullets. Red water washed on the backs of the faithful herd, as it gathered around the stricken one. We struggled for an hour to pick up the buoy and tie the harpoon line to the *Élie Monnier*.

A relatively small bottlenosed whale, heavily wounded, was tethered to the ship. We were out of sight of land, with fifteen hundred fathoms of water under the keel, and the whale herd diving and spouting around the ship. Tailliez and I entered the water to follow the harpoon line to the agonized animal.

The water was an exceptional clear torquoise blue. We fol-

BOTTLENOSE WHALES START A SHARK INCIDENT
We were cruising in the Élie Monnier *off Africa, when we spotted a shoal of bottlenose whales, 12 to 25 feet long. Dumas harpooned one. We went into the water, myself with a movie camera, and swam along the harpoon line toward the whale. Below the whale I saw an eight-foot shark of the* carcharhinus *family, but of an unusual type. His dorsal and pectoral fins were exceptionally large and rounded with white patches. (Continued.)*

lowed the line a few feet under the surface, and came upon the whale. Thin streams of blood jetted horizontally from the bullet holes. I swam toward three other bottlenoses. As I neared them, they turned up their flukes and sounded. It was the first time I had been under water to actually see them diving and I understood the old whaler's word, "sound." They did not dive obliquely as porpoises often do. They sped straight down, perfectly vertical. I followed them down a hundred feet. A fifteen-foot shark passed way below me, probably attracted by the whale's blood. Beyond sight was the deep scattering layer; down there a herd of leviathans grazed; more sharks roamed. Above in the sun's silvery light was Tailliez and a big whale dying. Reluctantly I returned to the ship.

Back on deck I changed into another lung and strapped a tablet of cupric acetate on an ankle and one on my belt. When this chemical dissolves in water it is supposed to repulse sharks. Dumas was to pass a noose over the whale's tail, while I filmed. Just after we went under he saw a big shark, but it was gone before I answered his shout. We swam under the keel of the ship and located the harpoon line.

A few lengths down the line in a depth of fifteen feet we sighted an eight-foot shark of a species we had never before seen. He was impressively neat, light gray, sleek, a real collector's item. A ten-inch fish with vertical black-and-white stripes accompanied him a few inches above his back, one of the famous pilot fish. We boldly swam toward the shark, confident that he would run as all the others had. He did not retreat. We drew within ten feet of him, and saw all around the shark an escort of tiny striped pilots three or four inches long.

They were not following him; they seemed part of him. A thumbnail of a pilot fish wriggled just ahead of the shark's snout, miraculously staying in place as the beast advanced. He probably found there a compressibility wave that held him. If he tumbled out of it, he would be hopelessly left behind. It was some time before we realized that the shark and his courtiers were not scared of us.

Sea legends hold that the shark has poor eyesight and pilot fish guide him to the prey, in order to take crumbs from his table. Scientists today tend to pooh-pooh the attribution of the pilot as a seeing-eye dog, although dissection has confirmed the low vision of sharks. Our experiences lead us to believe they probably see as well as we do.

The handsome gray was not apprehensive. I was happy to have such an opportunity to film a shark, although, as the first wonder passed, a sense of danger came to our hearts. Shark and company slowly circled us. I became the film director, making signs to Dumas, who was co-starred with the shark. Dumas obligingly swam in front of the beast and along behind it. He lingered at the tail and reached out his hand. He grasped the tip of the caudal fin, undecided about giving it a good pull. That would break the dreamy rhythm and make a good shot, but it might also bring the teeth snapping back at him. Dumas released the tail and pursued the shark round and round. I was whirling in the center of the game, busy framing Dumas. He was swimming as hard as he could to keep up with the almost motionless animal. The shark made no hostile move nor did he flee, but his hard little eyes were on us.

I tried to identify the species. The tail was quite asymmetrical, with an unusually long top, or heterocercal caudal fin. He had huge pectorals, and the large dorsal fin was rounded with a big white patch on it. In outline and marking he resembled no shark we had seen or studied.

Overleaf: The shark had an escort of three striped pilot fish, one over and one under his body and a tiny pilot riding the compressibility wave ahead of the shark's nose. Dumas, the shark, and I mingled underwater in shallow depth. We were hanging over a 12,000-foot abyss. Dumas swims down toward the shark and pilot fish. Second Overleaf: Dumas gets behind the shark and touches the caudal fin. Third overleaf: Above left, the beast turns, but not toward Dumas. He comes at me. Below left, the shark is ten feet away; pilot fish lose stations. Above right, I keep my finger on the camera button as he comes. Below right, the shark is two feet away. Then I bang his nose with the camera.

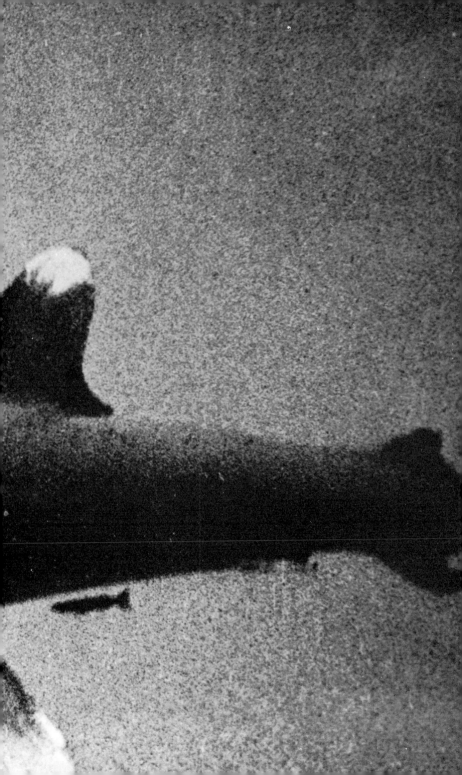

The shark had gradually led us down to sixty feet. Dumas pointed down. From the visibility limit of the abyss, two more sharks climbed toward us. They were fifteen-footers, slender, steel-blue animals with a more savage appearance. They leveled off below us. They carried no pilot fish.

Our old friend, the gray shark, was getting closer to us, tightening his slowly revolving cordon. But he still seemed manageable. He turned reliably in his clockwise prowl and the pilots held their stations. The blue pair from the abyss hung back, leaving the affair to the first comer. We revolved inside the ring, watching the gray, and tried to keep the blues located at the same time. We never found them in the same place twice.

Below the blue sharks there appeared great tunas with long fins. Perhaps they had been there since the beginning, but it was the first time we noticed them. Above us flying fish gamboled, adding a discordant touch of gaiety to what was becoming a tragedy for us. Dumas and I ransacked our memories for advices on how to frighten off sharks. *"Gesticulate wildly," said a lifeguard.* We flailed our arms. The gray did not falter. *"Give 'em a flood of bubbles," said a helmet diver.* Dumas waited until the shark had reached his nearest point and released a heavy exhalation. The shark did not react. *"Shout as loud as you can," said Hans Hass.* We hooted until our voices cracked. The shark appeared deaf. *"Cupric acetate tablets fastened to leg and belt will keep sharks away if you go into the drink," said an Air Force briefing officer.* Our friend swam through the copper-stained water without a wink. His cold, tranquil eye appraised us. He seemed to know what he wanted, and he was in no hurry.

A small dreadful thing occurred. The tiny pilot fish on the shark's snout tumbled off his station and wriggled to Dumas. It was a long journey for the little fellow, quite long enough for us to speculate on his purpose. The mite butterflied in front of Dumas's mask. Dumas shook his head as if to dodge a mosquito. The little pilot fluttered happily, moving with the mask, inside which Dumas focused in cross-eyed agony.

Instinctively I felt my comrade move close to me, and I saw his hand held out clutching his belt knife. Beyond the camera and the knife, the gray shark retreated some distance, turned, and glided at us head-on.

We did not believe in knifing sharks, but the final moment had come, when knife and camera were all we had. I had my hand on the camera button and it was running, without my knowledge that I was filming the oncoming beast. The flat snout grew larger and there was only the head. I was flooded with anger. With all my strength I thrust the camera and banged his muzzle. I felt the wash of a heavy body flashing past and the shark was twelve feet away, circling us as slowly as before, unharmed and expressionless. I thought, *Why in hell doesn't he go to the whale? The nice juicy whale. What did we ever do to him?*

The blue sharks now climbed up and joined us. Dumas and I decided to take a chance on the surface. We swam up and thrust our masks out of the water. The *Élie Monnier* was three hundred yards away, under the wind. We waved wildly and saw no reply from the ship. We believed that floating on the surface with one's head out of the water is the classic method of being eaten away. Hanging there, one's legs could be plucked like bananas. I looked down. The three sharks were rising toward us in a concerted attack.

We dived and faced them. The sharks resumed the circling maneuver. As long as we were a fathom or two down, they hesitated to approach. It would have been an excellent idea for us to navigate toward the ship. However, without land-marks, or a wrist compass, we could not follow course.

Dumas and I took a position with each man's head watching the other man's flippers, in the theory that the sharks preferred to strike at feet. Dumas made quick spurts to the surface to wave his arms for a few seconds. We evolved a system of taking turns for brief appeals on the surface, while the low man pulled his knees up against his chest and watched the sharks. A blue closed in on Dumas's feet while he was above. I yelled.

Dumas turned over and resolutely faced the shark. The beast broke off and went back to the circle. When we went up to look we were dizzy and disoriented from spinning around under water, and had to revolve our heads like a lighthouse beacon to find the *Élie Monnier*. We saw no evidence that our shipmates had spied us.

We were nearing exhaustion, and cold was claiming the outer layers of our bodies. I reckoned we had been down over a half hour. Any moment we expected the constriction of air in our mouthpieces, a sign that the air supply nears exhaustion. When it came, we would reach behind our backs and turn the emergency supply valve. There was five minutes' worth of air in the emergency ration. When that was gone, we could abandon our mouthpieces and make mask dives, holding our breath. That would quicken the pace, redouble the drain on our strength, and leave us facing tireless, indestructible creatures that never needed breath. The movements of the sharks grew agitated. They ran around us, working all their strong propulsive fins, turned down and disappeared. We could not believe it. Dumas and I stared at each other. A shadow fell across us. We looked up and saw the hull of the *Élie Monnier*'s launch. Our mates had seen our signals and had located our bubbles. The sharks ran when they saw the launch.

We flopped into the boat, weak and shaken. The crew were as distraught as we were. The ship had lost sight of our bubbles and drifted away. We could not believe what they told us; we had been in the water only twenty minutes. The camera was jammed by contact with the shark's nose.

On board the *Élie Monnier*, Dumas grabbed a rifle and jumped into the small boat to visit the whale. He found it faintly alive. We saw a brown body separate from the whale and speed away, a shark. Dumas rowed around to the whale's head and gave the *coup de grâce*, point-blank with a dum-dum bullet. The head sank with the mouth open, streaming bubbles from the blowhole. Sharks twisted in the red water, striking furiously at the whale. Dumas plunged his hands in the red froth and fastened a noose to the tail, which is what he had started out to do when we were diverted by our friend.

We hoisted the whale aboard and were impressed by the moon-shaped shark bites. The inch-thick leather of the whale had been scooped out cleanly, without rips, ten or fifteen pounds of blubber at a bite. The sharks had waited until we were cheated away from them before they struck the easy prey.

The whale became Surgeon Longet's biggest dissection. He swept his scalpel down the belly. Out on deck burst a slimy avalanche of undigested three-pound squids, many of them intact, almost alive. In the recesses of the stomach were thousands of black squid beaks. My mind leaped back to the fathogram of the deep scattering layer. The coincidence of the whale's lunch and the lines drawn on the fathogram may have been entirely fortuitous. It was not strict proof. But I could not dispel an unscientific picture of that dark gloaming of the scattering layer twelve hundred feet down, and whales crashing into a meadow writhing with a million arms of squids.

Standing for Dakar we met a porpoise herd. Dumas harpooned one in the back. It swam like a dog on a tether, surrounded by the pack. The mammals demonstrated a decided sense of solidarity. Save that the whale was now a porpoise, Dumas and Tailliez dived into a re-enactment of the previous drama. This time the dinghy carefully followed their air bubbles.

I watched the porpoise swimming on its leash like a bait goat a lion hunter has tied to a stake. The sharks went for the porpoise. It was cruelty to an animal but we were involved with a serious study of sharks, and had to carry it out.

The sharks circled the porpoise as they had circled us. We stood on deck remarking on the cowardice of sharks, beasts as powerful as anything on earth, indifferent to pain, and splendidly equipped as killers. Yet the brutes timidly waited to attack. Attack was too good a word for them. The porpoise had no weapons and he was dying in a circle of bullies.

At nightfall Dumas sent a *coup de grâce* into the porpoise.

When it was dead, a shark passed closely by the mammal, and left entrails in the water. The other sharks passed across the porpoise, muddying the sea with blood. There was no striking and biting. The sharks spooned away the solid flesh like warm butter, without interrupting their speed.

Sharks have never attacked us with resolution, unless the overtures of our friend and the two blues may be called pressing an attack. Without being at all certain, we suppose that sharks more boldly strike objects floating on the surface. It is there that the beast finds its usual meals, sick or injured fish and garbage thrown from ships. The sharks we have met took a long time surveying submerged men. A diver is an animal they may sense to be dangerous. Aqualung bubbles may also be a deterrent.

After seeing sharks swim on unshaken with harpoons through the head, deep spear gashes on the body and even after sharp explosions near their brains, we place no reliance in knives as defensive arms. We believe better protection is our "shark billy," a stout wooden staff four feet long, studded with nail tips at the business end. It is employed, somewhat in the manner of the lion tamer's chair, by thrusting the studs into the hide of an approaching shark. The nails keep the billy from sliding off the slippery leather, but do not penetrate far enough to irritate the animal. The diver may thus hold a shark at his proper distance. We carried shark billies on wrist thongs during hundreds of dives in the Red Sea, where sharks were commonplace. We have never had occasion to apply the billy, and it may prove to be merely another theoretical defense against the creature which has eluded man's understanding.

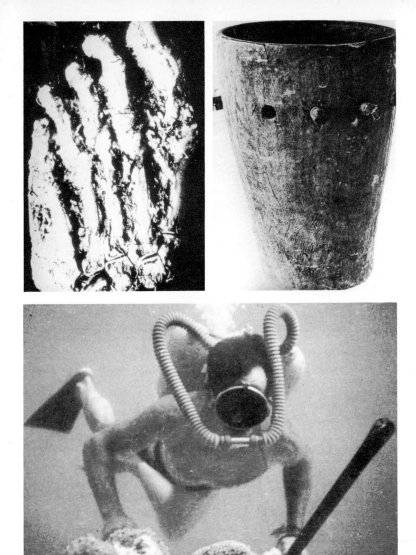

After our shark encounter we visited the Institut Français d'Afrique Noire *(French Institute of Black Africa) at Dakar and were shown these official pictures of objects found in sharks' stomachs: a wooden tom-tom and the remains of a human foot. (Photo by Labitte.) Below, we sometimes carry shark billies, wooden clubs to poke sharks on the nose.*

Chapter Thirteen **Beyond the Barrier**

Most of our dives have had a specific purpose—wreck exploration, de-mining, or experimenting in physiology, for instance. But occasionally we were able to steal hours of dawdling inside the sea, where a man could invite his senses to the nuances of color and light, listen for the lonely creaks of the ocean and finger the water like a voluptuary. Then one realized the privilege of crossing the barrier, that molecular tissue which is actually a wall between elements. If it was difficult for men to break through the wall, it was more difficult for fish, which told us with their brief awkward leaps into the air how alien were air and water.

One of the greatest joys of sea bathing, perhaps not realized by many, is that water relieves the everyday burden of gravity. Human beings, and other vertebrates which live in air, expend a great deal of energy in simply holding themselves up. The sea does it for you. The air in one's lungs provides positive buoyancy, and heaviness falls from the limbs, bestowing a relaxation no bed can offer.

It is a popular notion that fat people float better than thin ones. Fatty tissues weight slightly less than muscles, yet tests that we made on stout and thin people showed that the heavier ones had no buoyancy advantage. We believe the contradiction is due to the tendency of stout people to have less lung development than lean folk. Lung ballast is decisive in floatability. As a matter of fact a novice usually requires more belt weights than an experienced diver of like displacement. The beginner, chilled and apprehensive, involuntarily inflates his lungs too much and needs more ballast to be equilibrated. After a few dives he breathes normally and finds that he is overweighted.

Then he learns the wide possibilities of adjusting air ballast by his own breathing discipline, a factor that has a range of six to twelve pounds on his displacement.

In his world without weight, the diver orients himself to the strange behavior of inanimate objects. When a hammer breaks, the head sinks and the handle rises. Underwater tools have to be ballasted so they will not fly off in all directions. Knife blades are counterpoised with cork. Seventy-pound cameras have enough air in the housing to make them weightless. The balance must be very sharp, because carrying a single tool which is out of equilibrium can disturb a diver's own ballast. At the beginning of a dive the compressed air in one aqualung cylinder weighs three pounds. As it is consumed the diver weighs less with each breath. When the air is gone the cylinder exerts a three-pound ascension force. The perfectly adjusted dive begins with the diver slightly overweighted, a logical situation, since he wishes to sink. He is slightly underweighted when the dive ends, a logical situation since he must return to the surface.

Before a camera dive I look like a beast of burden staggering into the water with a forty-five-pound triple aqualung strapped on my back, four pounds of lead on my belt, and the weight of knife, pressurized watch, depth gauge, and compass on my wrists, and perhaps a four-foot shark billy on a wrist thong. It is a relief to float my burdens in the sea and receive the seventy-pound Bathygraf cinécamera, either from two men on deck, or lowered by davit. Fully borne, myself and my gear weigh two hundred sixty-five pounds. Immediately under water I weigh a mere pound or so—the deliberate overweight—and swim head down with wonderful ease.

Weight has been suppressed, but not inertia. It takes several strong kicks of the fins to get the caravan under way and I glide on after ceasing to kick. It is unwise, if not impossible, for an ill-designed hydrofoil like a man to swim rapidly under heavy water. It is best to let the element set the speed, a languorous journey, a slow-motion flight true to the physics of the environment.

As one swims down, pressure increases steadily and rapidly. Each foot of descent adds nearly a half pound of pressure per square inch on the body. Aside from the "wedging" sensation in the ears, which is overcome by swallowing, one feels no subjective reaction to pressure. Human tissue is almost incompressible. We have swum without armor in pressures that have cracked submarine hulls. They did not have the necessary counter-pressure supporting them on the inside.

A man on land bears an atmospheric pressure of several tons on the surface of his body without noticing it. The oceanic fluid doubles atmospheric pressure thirty-three feet down. At sixty-six feet pressure is tripled; it is fourfold at ninety-nine feet, and so on down, in multiples of thirty-three feet.

There are animals living thirty thousand feet under the sea, where pressure is exerted on their bodies at the rate of seven tons per square inch. If pressure were the only problem for men in the sea, we could descend to at least two thousand feet without armor. However, indirect effects of pressure would stop a man far short of that depth. The absorption of vast amounts of gases in his tissues and the inability to pass off carbon dioxide would limit such a theoretical dive.

Indeed pure pressure changes become progressively easier on a diver the deeper he goes. A man who goes up and down several times in the thirty-three-foot layer experiences pain and exhaustion because he is doubling his external pressure every time he reaches thirty-three feet. But his colleague who goes further down meets less radical adjustments. Between thirty-three and sixty-six feet he experiences only half as much change as in the first layer. In the next thirty-three feet pressure is increased by only a third, and by a fourth in the next strata. Be-

---

Overleaf, five menfish sound for the bottom. Frédéric Dumas, Jean Pinard, Guy Morandière, Philippe Tailliez, and Quartermaster Georges are 60 feet down. I am down 75 feet, pointing the camera toward the surface.

tween one hundred and thirty-two feet and one hundred and sixty-five feet, the weight of water increases by a fifth. We have a rule of thumb that a person who is physically able to stand a dive of ten yards can most likely go to two hundred feet without physical failure. The critical area is at the top.

Because of this fact, the most dangerous area for the helmet diver is the superficial layer. The helmet and the upper part of the suit imprison a large air bubble, which is acutely sensitive to pressure variations. As he passes through the decisive top layer, the helmet diver must control his air supply carefully to avoid "the squeeze" and "ballooning." A ballooned diver is one who admits too much air to his suit. In the danger zone, it suddenly swells, blows the suit up like a blimp and shoots him helplessly to the surface, in danger of nitrogen embolism from the bends.

"The squeeze" is the opposite effect, the lack of counterpressure in the helmet and lungs. The helmet becomes a monstrous *ventouse*, the old-fashioned doctor's cupping glass which was applied to the chest of a coughing person. In the Undersea Research Group we called "the squeeze" *coup de ventouse*, a blow of the cupping glass. A helmet diver whose air pipe has failed is usually killed by *coup de ventouse*. If the nonreturn valve in his air pipe does not hold, his fate is horrible. By the suction of the air pipe his flesh is stripped away in rags which stream up the pipe, leaving a skeleton in a rubber shroud to be raised to the tender.

Breathing under water is an attractive notion. As a boy in Alsace I read a wonderful story of a hero who hid from villains by breathing through a hollow reed from a river bottom. (Folklorists could probably find the device in all languages.) I put a length of garden hose through a block of cork, took the breathing end in my mouth, clutched a stone, and jumped into the swimming pool. I couldn't suck a breath. I abandoned my hose and stone and made off, frantically, for the surface. Boys thereby gather cynical impressions about storybooks. The author of my book, like others who write diving yarns, had never been on a river bottom with a straw in his mouth.

A few feet down the pressure exceeds the ability of human respiratory muscles to pull down surface air. A man with extraordinary lungs might draw air six feet down for a few minutes, but breathing surface air a foot down is tiring for most of us. A schnorkel tube reaching up six inches can fatigue a person who feasts his eyes too long in the sea.

Bathers like warm seas and so do divers. But, alas, the pleasure is mitigated when one goes down. The warmest water of the Mediterranean is in August, but it is a thin surface layer. Then one enters cooler water, but still fairly comfortable. In June and November the moderate zone extends to forty-five feet. In July, August and October it goes down to one hundred and twenty-feet. September is the best time—the sea is charitable down to two hundred feet.

Beneath the moderate layer one hits cold with a physical crash—a temperature of fifty-two degrees Fahrenheit. The warm and cold layers meet with the precision of wood veneers. There is no transition. One can hang in the moderate zone and poke a finger into the cold and feel it as sharply as one sticks an exploratory toe in the sea on the season's first dip. We often hesitate, shivering at the thought, screwing up courage to take the plunge. Once inside the cold layer the skin is soon anesthetized and the diver may comfort himself thinking of his return across the knife edge of warmer water. Shrewd submarine skippers have floated their vessels on the boundary. The cold water is a little heavier. The submariner trims his ballast slightly heavy for the warm water and a bit lighter for the cold water, stops his engines, and the sub hangs there.

The mistral that came while we were exploring the *Dalton* introduced us to the dynamics of warm and cold layers. We were enjoying moderate water to one hundred and twenty feet,

Overleaf, Tailliez is swimming in the reefs off Cassis when he discovers a cave 90 feet down. He calls it "Ali Baba Cave" because of its rich fish, sponges and corals.

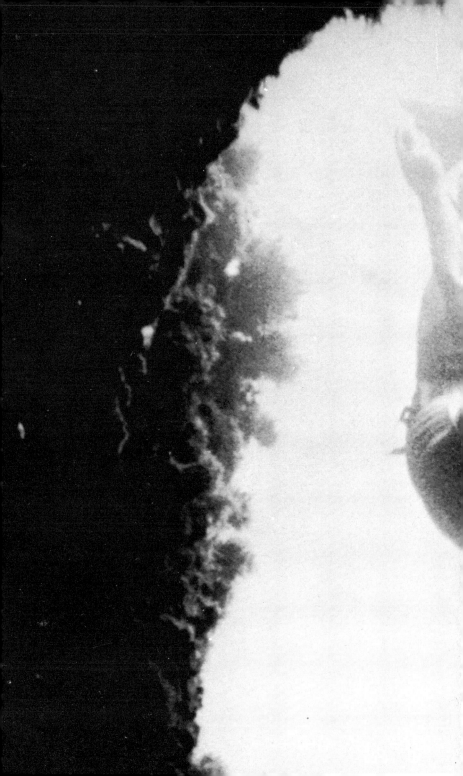

when the north-northwest storm blew down. On the gale's first day the cold strata rose to eighty feet, and arrived at forty feet the next day. On the third day the pelagic refrigerant reached the surface and an even fifty-two degrees Fahrenheit prevailed from top to bottom. It was proof that the mistral did not cool the surface water, but blew it away and cold water from the depths replaced it. When the wind died, warm water gradually returned from the sea, forcing the frigid level deeper each day. That is how we came to find icy water in the holds of the *Dalton* when the sea all around was comfortable.

We dived to Tailliez's "Ali Baba Cave," ninety feet down off Cassis on a day when cold prevailed from floor to waves. Inside the cave we found big tubs of delicious tepid water the wind had not been able to uproot and blow away.

Anyone who has gone swimming in the rain knows the silly feeling of "dryness" under the surface and the hesitation to come out lest one get wet. A diver looking up toward rain sees a million tiny moving peaks speckling the water. Fresh water, slowly and stubbornly mixing with the salt sea, creates an area of optical distortion in the superficial layer, like heat rays dancing above hot earth. In coastal waters during rainstorms we have seen an extraordinary excitement among the fish. They go crazy in the rain. The little ones explode in all directions. From the bottom come sedentary sars, climbing and diving, moved to sensational exertions. Mullets and bass jig frantically under the boiling carpet of rain. They stand on their tails with mouths open as if sucking the fresh water. Rainy days in the sea are wild celebrations.

The sea is a most silent world. I say this deliberately on long accumulated evidence and aware that wide publicity has recently been made on the noises of the sea. Hydrophones have recorded clamors that have been sold as phonographic curiosa, but the recordings have been grossly amplified. It is not the reality of the sea as we have known it with naked ears. There are noises under water, very interesting ones that the sea transmits exceptionally well, but a diver does not hear boiler factories.

An undersea sound is so rare that one attaches great importance to it. The creatures of the sea express fear, pain and joy without audible comment. The old round of life and death passes silently, save among the mammals—whales and porpoises. The sea is unaffected by man's occasional uproars of dynamite and ship's engines. It is a silent jungle, in which the diver's sounds are keenly heard—the soft roar of exhalations, the lisp of incoming air and the hoots of a comrade. One's hunting companion may be hundreds of yards away out of sight, but his missed harpoons may be clearly heard clanging on the rocks, and when he returns one may taunt him by holding up a finger for each shot he missed.

Attentive ears may occasionally perceive a remote creaking sound, especially if the breath is held for a moment. The hydrophone can, of course, swell this faint sound to a din, helpful for analysis, but not the way it sounds to the submerged ear. We have not been able to adduce a theory to explain the creaking sounds. Syrian fishermen select fishing grounds by putting their heads down into their boats to the focal point of the sound shell that is formed by the hull. Where they hear creaking sounds they cast nets. They believe that the sound somehow emanates from rocks below, and rocks mean fish pasturage. Some marine biologists suppose the creaking sound comes from thick thousands of tiny shrimps, scraping pincers in concert. Such a shrimp in a specimen jar will transmit audible snaps. But the Syrians net fish, not shrimps. When we have dived into creaking areas we have never found a single shrimp. The distant rustle seems stronger in calm seas after a storm, but this is not always the case. The more we experience the sea, the less certain we are of conclusions.

Some fish can croak like frogs. At Dakar I swam in a loud orchestration of these monotonous animals. Whales, porpoises, croakers and whatever makes the creaking noise are the only exceptions we know to the silence of the sea.

Some fish have internal ears with otoliths, or ear stones, which make attractice necklaces called "lucky stones." But fish show little or no reaction to noises. The evidence is that they are much more responsive to nonaudible vibrations. They have a sensitive lateral line along their flanks which is, in effect, the organ of a sixth sense. As a fish undulates, the lateral receiver probably establishes its main sense of being. We think the lateral line can detect pressure waves, such as those generated by a struggling creature at a great distance. We have noticed that hooting at fish does not perturb them, but pressure waves generated by rubber foot fins seem to have a distinct influence. To approach fish we move our legs in a liquid sluggish stroke, expressing a peaceful intention. A nervous or rapid kick will empty the area of fish, even those behind rocks which cannot see us. The alarm spreads in successive explosions; one small fleeing creature is enough to panic the others. The water trembles with emergency and fish far from sight receive the silent warning.

It has become second nature to swim unobtrusively among them. We will pass casually through a landscape where all sorts of fish are placidly enjoying life and showing us the full measure of acceptance. Then, without an untoward move on our part, the area will be deserted of all fish. What portent removes hundreds of creatures, silently and at once? Were porpoises beating up pressure waves out of sight, or were hungry dentex marauding off in the mists? All we know, hanging in the abandoned space, is that an unhearable raid siren has sent all but us to shelter. We feel like deaf men. With all senses attuned to the sea, we are still without the sixth sense, perhaps the most important of all in undersea existence.

At Dakar I was diving in water where sharks ranged peacefully among hundreds of tempting red porgies, unwary of the

Dumas ascends a 120-foot rock chimney beneath the sea. Overleaf, to the undersea explorer the canyons are out-of-scale and unbelievable; to rise along the wall is an impossible dream.

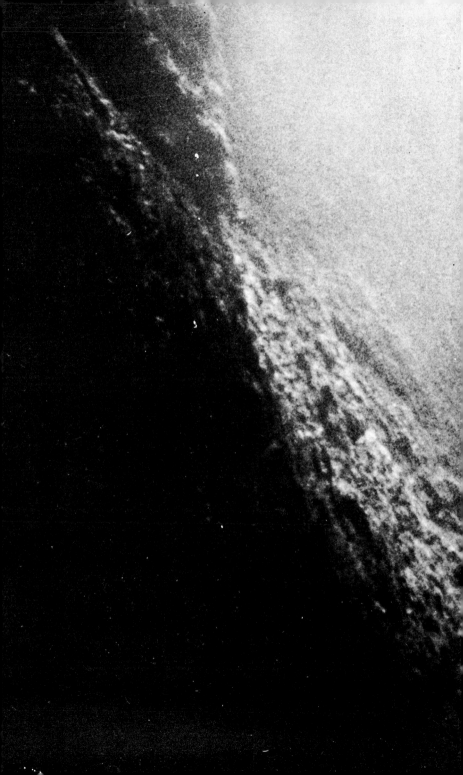

predators. I returned to the boat and threw in a fishing line and hooked several porgies. The sharks snapped them in two before I could boat them. I think perhaps the struggle of the hooked fish transmitted vibrations that told the sharks there was easy prey available, animals in distress. In tropical waters we have used dynamite to rally sharks. I doubt whether the explosion is anything more than a dull, insignificant noise to them, but they answer the pressure waves of the fluttering fish that have been injured near the burst.

On the Azure Coast there are vertical reefs two hundred feet deep. Going straight down one of these walls is an unusual excursion into the variety of the sea and its abrupt changes of environment. Mountain climbers, like our friend, Marcel Ichac, who have gone down the reefs with us, are surprised at the changes. Going up a mountain one struggles through miles of foothills, through extended zones of trees to the snow-line, to the tree-line, and into the thin air. On the reef the changes are rapid, almost bewildering, from one zone to another. The top

ten fathoms, lighted by sunny lace from the surface, are populated with nervous darting fish. Then one enters a strange country upon which dusk has fallen at noon, an autumnal clime with insalubrious air that makes the head heavy, like that of a person doomed to live in a smoggy industrial town.

Gliding down the rock façade one looks back at the world where summer shines. Then one comes to the cold layer and grows tense for the leap into winter. Inside the dull dark cold one forgets the sun. One forgets a lot. The ears no longer announce pressure changes and the air tastes like pennies. An introspective calm rules there. The green mossy rocks are replaced by Gothic stones, pierced, cusped and enfinialed. Each vault and arcade of the bottom rocks is a little world with a sandy beach and a tableau of fish.

Deeper down are miniature blue trees with white blossoms. These are the real coral, the semiprecious *corallium rubrum* in brittle limestone fantasies of form. For centuries coral was commercially dredged in the Mediterranean with "coral crosses," a type of wooden drag that smashed down the trees and recovered a few branches. The once-thick trees on the floor that may have taken hundreds of years to grow, are no more. The surviving coral grows below twenty fathoms in protected recesses and grottoes, accumulating from the ceiling like stalactites. It may be gathered only by divers.

A diver entering a coral cave must be aware of its appearance in the sea's deceiving color filter. The coral branches appear blue-black. They are covered with pale blossoms that retract and disappear when disturbed. Red coral is out of fashion at the moment and sells for about ten dollars a pound.

In the zone of red coral black-striped lobster horns protrude from the lacunae of the reef. When a diver's hand comes near, the lobster stirs with a dry grating sound. On the rocks are living tumors and growths resembling udders, long fleshy threads, chalice-shaped formations, and forms like mushrooms.

Dr. Devilla, surgeon of the Undersea Research Group, dives for his dinner and comes up with a 19-pound lobster.

Objects may no longer be distinguished by color, although there are supernatural colors to them—the violet of wine dregs, blue-blacks, yellowish-greens; all muted and grayed, but somehow vibrant.

Now at the base of the reef, the sand begins, bare and monotonously receding into the floor. There, on the border of life, nothing grows or crawls. One moves automatically without brain directives. In the recesses of the brain, one revives an old notion—return to the surface. The drugged state disappears on the rise along the wall, the departure from a discolored land, a country that has never shown its real face.

Fish do not like to go up or down, but swim on a chosen level of the reef, like tenants of a certain floor of a skyscraper. The ground-floor occupants, wrasses, groupers and Spanish bream, rarely venture to upper stories of the cliff dwelling. The dentex pace back and forth just above the sandy fields. The sars pass in and out of the rocks with a busy and determined air. The wrasses are slow and seem bored. The Spanish bream are slower still. They hang against the cliff sucking the rocks like lollipops. Higher and away from the reef tower, the pelagic fish roam, but they too seem to prefer a given stratum and rarely swim up or down. Fish do not like the effort of making pressure changes.

Fish seem to glide forever as long as one does not startle them. What do fish do all day long? Most of the time they swim. We have rarely witnessed fish feeding. The sar is sometimes found along a rock, browsing with its goatish teeth on clinging sea urchins. It methodically clips away the urchin's brittle spines, spits them out, and chews down until it is able to cut open the carapace and reach the meal inside. *Rouquiers** eat all the time. They gobble invisible tidbits from the soil, or stand vertically, blowing small dust clouds and swallowing them. The fussy mullet rustles across the rocks, sucking weeds with its thick white lips, cleaning off the fish eggs and spores. Sea bream graze by the hundreds on the ocean prairies. When

* There seems to be no English or American equivalent for the name of this fish.

our presence disturbs them they relieve themselves in immense green clouds and depart.

We have waited for years to witness the meal of a carnivorous bass, dentex, conger or moray, and have never seen it. We know only from surface observation that the predators have two strict daily mealtimes, in the morning and evening, as regularly observed as a boarding-school dinner gong. The vast shoals of sprats, sardines, or needlefish, living near the surface, are savagely attacked from below. The sea boils and the air flickers with the hail of little bodies, leaping out and falling back. Sea birds join the massacre from the other side of the barrier, diving and flaunting the sparkling prey in their bills. When we dive in the banquet stops. We see the big ones roving below, waiting for us to leave. The small ones find a moment of surcease near the surface. Fish will not feed near divers. The war for food lasts a half hour, then a truce comes and eaters and those to be eaten tomorrow mingle sociably again in the quiet flood.

Eagerly as we have sought to observe carnivorous feeding, so have we awaited almost as vainly for the mating drama. Mullets are the most shameless. They breed in September in the warm shore waters of the Mediterranean. The females stroll up and down with composure while the excited males flutter around, rubbing themselves feverishly against their mates. At that season the majestic Spanish bream forget their lonely arrogance and aggregate in incredible swarms, pressed so closely against each other there is scarcely room to slide. No two fish hold the same position in the amorous mêlée, quite unlike the normal formations of a school.

Fish have different ways of showing their curiosity. Often while swimming along we will turn back abruptly and see the muzzles of echelons of creatures following us with avid interest. The dentex gives us a passing glance of contempt. The sea bass approaches us, investigates and swims away. The liche feigns indifference, but closes in for a better look and is quickly satisfied.

Not so with the merou. The grouper is the ocean's scholar, sincerely interested in our species. It approaches and looks at us with large, touching eyes, full of puzzlement, and stays to survey us. The merou probably attains a hundred pounds. We have speared and weighed individuals of fifty-five pounds and have seen others which seemed twice that weight. The creature is a cousin of the tropical jewfish, which grows as big as five hundred pounds. The merou lives near the coast in thirty feet of clouded turbulent water, close to the rock forts it holds so stubbornly. There are a few radical merous who have taken caves further up on the shelf in ten-foot depths. They are suspicious individualists who rarely emerge, have a clear conception of danger, and grow old peering out their doors.

They are the most inquisitive animals we have found in the sea. In virgin territory merous swim out of their holes and come great distances to see us. They sit below and look up full in our faces. With their big pectorals spread like the wings of baroque angels, they stare sanctimoniously at us. When we move they shake themselves, and leap to new vantage points. When at last they go back home, they watch us from their doors and run to a window to see us depart.

When the tiny black pomfrets with forked tails, half the size of a goldfish, throng around the merous, the groupers stare through at us like veiled women. If we dive into the screen of pomfrets, it breaks like a window glass and the merou vanishes. A hundred feet down the fish apparently do not associate us with the surface. In the sad bluish gloom one is accepted in the jungle and its inhabitants have no fear, merely curiosity toward the extraordinary animal with a mania for spreading bubbles.

The merou eats everything in the path of its huge open mouth. In go octopi and the stones they may be clinging to, whole cuttlefish in their shells, thorny sea spiders, lobsters and entire fish. If the merou accidentally swallows a fishing hook, it usually severs the line. One of Dumas's merous had two fish

Above, Dumas finds that a grouper's mouth opens as wide as its circumference. Below, Dumas chases a grouper around the corner.

hooks in its stomach, the metal encysted with age. The merou has a chameleon's talent. Mostly they are reddish brown. They can put on a marbleized pattern or dark stripes. Once we found a white one flat on the sand. We thought it was the pallor of death and decay, but the ghost stirred up, turned brown, and made off.

One morning we were swimming across a large fissure fourteen fathoms down. We halted and lay in the water, looking down at a group of twenty- to thirty-pound adolescent merous. They swam straight up toward us and then turned over and glided down like children on a slide. Below them were a dozen larger individuals moving in a preoccupied figure. One of the merous turned white. The others paraded by it closely. One stopped beside the albino and itself turned white. Then both uncolored animals rubbed themselves slowly against each other, perhaps in an act of love. We stared, unable to comprehend the sight. What was the ceremony in the dim rocks? It was as strange as the elephant dance that little Toomai saw.

The merou has a special and old familiar place in our undersea experience. We feel sure that we could tame one, by training that generous curiosity toward becoming a pet.

Chapter Fourteen **Where Blood Flows Green**

TROPICAL peoples have always known that a man has to go into the sea to earn its bounty, but until some remote Polynesian put two pieces of glass in a watertight spectacle frame, men were almost blind under water. The refraction index of water against a naked eyeball nullifies the curvature of the cornea. Instead of converging on the retina, images are formed behind it in the blue seen by hopelessly farsighted persons.

Through the mask the diver sees objects larger than they are. Things seem one-fourth nearer than their actual distance, a deceitful perspective caused by the refraction of light passing from water to air through the glass plate. On my first dive I reached for objects, saw my hand fall short and was dismayed at my shrunken flipper of an arm. The enlargement factor is helpful in telling fish stories. A six-foot shark expands to nine feet without much imaginative effort. It takes practice to automatically correct distance and size.

Sometimes when diving with Dumas I have stalked him, pretending to be a shark. It was easy to hang back and hide in his blind fields. When Dumas sensed there was a shark after him, he would use every trick to bring me into sight. It was not difficult to avoid detection by watching his movements and altering mine. If a man could do that to such a skilled swimmer as Dumas, it posed grim anticipations of what a half-intelligent man-eater could do before it was seen.

A new day begins in the sea with the faintest of change in light. The predawn glow suffuses its light into the dark depths, but when the sun itself appears there is no burst of light, because the low sunbeams glance off the surface. The sun's direct rays do not strike into the gulf until it is riding overhead at noon. In the evening the sea light fades gradually and there is no sunset. Daylight dwindles down to starlight, or moonlight, and dark.

Sunlight penetrating the sea loses intensity as its energy is transformed to heat by absorption. Light is further diffused by particles suspended in the water, mud, sand and plankton and even by the water molecule itself. The particles are like motes in a sunbeam, reducing visibility and scattering the light supply before it can reach the great depths. The voids are black, like interplanetary space where no floating particles reflect the light of the sun.

In clear water it seems very dark a hundred feet down, but when the diver reaches the bottom it becomes half-bright again because the light reflects from the floor, the phenomenon we found in the aftercastle of the *Dalton*.

Dumas accustoms his eyes to the darkness before entering a cave 100 feet down.

At the three-hundred-foot limit of aqualung diving there is usually enough light to work by and often to expose black-and-white photographs. Dr. William Beebe and others have measured light penetration to fifteen hundred feet.

Not only does water transparency vary from place to place, but from level to level. Once we made a dive to a submerged rock needle in the Mediterranean. The water was so turbid we could see only a few yards. Two fathoms down the water suddenly opened in a clear layer, going down some distance. Then came a fifteen-foot layer of milky water with about five-foot visibility. Under the milk was a clear world all the way to the bottom. The fish were lively and abundant in the shaded but very clear depths. Above us the foggy layer looked like a low overcast on a rainy day. Often on deep dives we pass through alternate layers of turbidity and clarity, puzzling in their dynamics.

Indeed, a given layer will sometimes change clarity before one's eyes. I have seen clear water grow dim, with no apparent current to shift the scenery, and have seen murk dissolve just as mysteriously. In the open sea our experience has been that the most turbid layer is on the surface in spring and fall, but again in those seasons we have found it deeper down under a thick layer of limpid water.

Turbidity near shore can be caused, of course, by silt particles emptied from rivers. But at sea, beyond this influence, opacity is mainly due to innumerable micro-organisms. In the late spring the water is saturated with algae, tiny one-celled and multi-celled animals, spores and eggs, minuscule crustaceans, larvae, living filaments and pulsating blobs of gelatin. These culture broths can reduce visibility to fifteen feet. Swimming through the soup may give the diver a germophobia. He does not like to have the multitudes of little creatures sliding along his skin. Indeed there is harm in the oceanic mites. One feels pin pricks and sharp burns in unexpected places, most painfully on the lips. It is just as well that one's eyes are behind glass.

People who describe the enchanting riot of color in tropical reef fairylands are talking about the environment down to perhaps twenty-five feet. Below that, even in sunflooded tropical shallows, one can see only about half the real color values. The sea is a bluing agent.

The color metamorphosis of the sea was studied by the Undersea Research Group. We took down color charts with squares of pure red, blue, yellow, green, purple and orange, together with a range of grays from white to black, and photographed the chart at various levels down to the twilight zone. At fifteen feet red turned pink, and at forty feet became virtually black. There also orange disappeared. At a hundred and twenty feet yellow began to turn green, and everything was expressed in almost monochromatic colors. Ultraviolet penetrated quite deeply, while infrared rays were totally absorbed in inches of water.

One time we were hunting under the isolated rocks of La Cassidaigne. Twenty fathoms down Didi shot an eighty-pound liche. The harpoon entered behind the head but missed the spinal column. The animal was well hooked but full of fight. It swam away, towing Didi on his thirty-foot line. When it went down, he held himself crosswise to cause a drag. When the fish climbed, Didi streamlined himself behind it and kicked his flippers to encourage ascent.

The liche fought on, untiringly. It was still struggling as we drew near the depletion of our air supply. Dumas hauled himself forward on the line. The fish circled him at a fast, wobbling speed, and Dumas spun with it to avoid being wound up in the line the way Tashtego was lashed to Moby Dick. Dumas hauled in the last feet of cord, and got a grip on the harpoon shaft. He flashed his belt dagger and plunged it into the heart of the big fish. A thick puff of blood stained the water.

The blood was green. Stupefied by the sight, I swam close and stared at the mortal stream pumping from the heart. It was the color of emeralds. Dumas and I looked at each other wildly. We had swum among the great liches as aqualungers and taken them on goggle dives, but we never knew there was a type with green blood. Flourishing his astounding trophy on the harpoon, Didi led the way to the surface. At fifty-five feet the blood turned dark brown. At twenty feet it was pink. On the surface it flowed red.

Once I cut my hand seriously one hundred and fifty feet down and saw my own blood flow green. I was already feeling a slight attack of rapture. To my half-hallucinated brain, green blood seemed like a clever trick of the sea. I thought of the liche and managed to convince myself that my blood was really red.

In 1948 we took light into the twilight zone. At high noon in clear water Dumas went down with an electric lamp, as powerful as a movie "sunlight," with a cord to the surface. Although our eyes could well distinguish the blue forms of the twilight, we wanted to see the real colors of the place.

Didi trained his reflector on the reef wall one hundred and sixty feet down. He snapped the light on. What an explosion!

The beam exposed a dazzling harlequinade of color dominated by sensational reds and oranges, as opulent as a Matisse. The living hues of the twilight zone appeared for the first time since the creation of the world. We swam around hastily, feasting our eyes. The fish themselves could never have seen this before. Why were these rich colors placed where they could not be seen? Why were the colors of the deep the reds that were first to be filtered out in the top layers? What are the colors further down, where no light has ever penetrated?

It set us off in a technical drive to make color photography in the blue zone that begins roughly one hundred and fifty feet down. For ten years we had been working in black-and-white cinéfilm. Photography under water goes back further than one realizes.

One day we came upon a rare book called *La Photographie Sous-Marine (Undersea Photography)* by Louis Boutan, published in 1900. He described six years of experiment in undersea photography in the days when pictures were made on clumsy wet glass plates. Boutan had made the first undersea photos in the Bay of Banyuls-sur-Mer in 1893!

Tailliez made our first underwater movies with a 9.5 mm. Pathé camera sealed in a two-quart fruit jar. The American, J. E. Williamson antedated us on this achievement. There is no doubt that he made the first underwater movies in 1914.

When we began filming under water we did not encounter any optical problems. The movies came out in excellent focus, determined simply by our sense of distance. We had forgotten about refraction of light passing through the porthole from water to air. Later our shots went out of focus. The same operator and the same camera failed to get clear images. It was a disheartening experience, which we analyzed and solved, not with optics, but with psychology. We had instinctively focused on what seemed the true distance and the camera faithfully recorded what we saw. Then we got too smart. We learned to correct the distance automatically in our minds and focused on the true distance. That blurred the shots because the lens could not make that mental correction. When we went back to estimating focal length by what it *seemed* to be, the pictures were as sharp as before.

Underwater cinematography with a hand-held camera was a revelation. The camera followed the imagination. Our cameras, still or movie, hang from shafts with two pistol grips like submachine guns. The operator holds them ahead of himself, stalking his prey. The embrace of water allows shots that in a studio require complex booms, turrets and carts. In air, hand-held cameras inevitably jitter. The water smooths things out for superb traveling shots toward an object, steady panoramic shots and involved three-dimensional movements that would take a day to rehearse on a professional studio boom.

We never use viewfinders in undersea work. The camera is pointed ahead, like the weapon of the hunters we once were and we "fire" at the target without using sights. The thrust of the cameraman's body, his eyes, and the lens are co-ordinated.

Our first diving film was sharply lighted by sunlight in

shallow water. When we went deeper with the lung we found that we could continue exposing good black-and-white images further down. In 1946 we made clear movie sequences two hundred and ten feet down at noon in July without artificial light. The exposure was one-fiftieth of a second at an opening of f/2. Since all gradations of light could be found between that depth and the surface, we installed underwater aperture controls to capture the modulations of gray. In 1948 we discovered that color movies could be made in natural light at surprising depths, when we filmed the divers at work on the argosy of Mahdia in one hundred and twenty-seven feet.

We did not dive to make movies. We made movies to record dives. Most of the seventy thousand feet of underwater film we have made remains in our archives. Without film we would have never been able to secure Navy authorization to form the Research Group. The movies have been directly instrumental in launching our oceanographic expeditions. But, strangely enough, scientific purposes are better served by still photographs in color. After making motion pictures for a decade, we

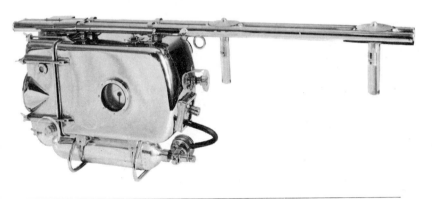

Our latest undersea movie camera, the Bathygraf. The pistol grips turn to control aperture and focus. It weighs minus one pound underwater (70 pounds in air). It is automatically pressurized by the aqualung cylinder under camera. It runs on a tiny silver-potassium battery.

set up a parallel still photographic project. In many ways stills are more difficult than movies.

In 1926 Dr. W. H. Longley and Charles Martin of the National Geographic Society made the first underwater color photographs, in the Dry Tortugas, using a floating reflector and magnesium powder which lighted the water to fifteen feet. But no surface flash could reach the twilight zone. We had to introduce artificial light in the blue layer—a hundred and fifty feet down.

Francois Girardot, the Parisian specialist in underwater photographic equipment, who was by now constructing our cameras, built a Rolleiflex unit on our favorite principle of the submachine gun shaft with pistol grip controls, and self-pressurization. Since existing electronic flash equipment was too weak, in relation to its size and weight, we constructed a flash-bulb reflector that held eight of the most powerful flash bulbs known, a "slow peak" type, each of which would deliver five million lumens. At night on land the bulb would light color subjects fifty feet away. In the opaque sea it had only a six-foot radius.

The camera had two flash extensions which were hand-held by divers. Selector knobs on the reflectors allowed us to fire one, two, four or eight bulbs simultaneously. The top load flashed over forty million lumens, a light that, except perhaps for the inner burst of an atomic bomb, had not been unloosed in a small area. This power could illumine at most a fifteen-foot radius of water. We found that the bulbs would go as deep as we could without crushing. We called the effort "Expedition Flash" and went into the Mediterranean to make the first color work in appreciable depth.

Jean Beltran and Jacques Ertaud, holding the reflectors attached to the camera by thirty-foot cables, joined me underwater and we swam down abreast. The cables were supported by small buoys so they would arch overhead, stay out of the camera field, and have less likelihood of snagging on rocks. Dumas swam ahead to twenty-five fathoms and selected a coral grotto as the subject. He was to appear in the pictures to give them scale.

The crew arrived in a dark place in which we were barely able to see Didi's bright foam-rubber jerkin against the blue reef. He placed a color chart against the wall to check the values of the photograph. No one knew what combination of color film and flash bulb would give true rendering. Submarine color photography had no prior laws. We had to find them out.

Suspended in the sea, Ertaud and Beltran trained their reflectors on Didi, in the familiar studio lighting scheme of one

Dumas down on the sea floor with a movie camera, from a color stereophotograph by our skilled diving companion, Jean de Wouters, who built his own underwater stereocamera. (Photo by Jean de Wouters d'Oplinter.)

beam close to the subject and one further away overhead for general diffused lighting. I fired. We saw an outrage of color, so quickly gone that we had no distinct idea of forms, but lay blinded in the darkness, trying to "swallow" the after-images that flared in our retinas. It took some time to recover from the terrible light.

Didi moved to another scene and we set up again. This time the bulbs did not fire. We returned to the surface. The bulbs were intact; they had successfully resisted pressure nearly five times that of the atmosphere. But they would not fire on the surface, either. At the laboratory we found that the first shot was a blank. Water had seeped into the lamp sockets.

The only way to salvage the project was to use watertight reflectors, something we had of course considered at the beginning, and had tried to avoid because of the heavy expense and time involved. Fragile naked bulbs would withstand pressure, but once they were hermetically sealed behind a glass porthole, the reflector would have to be pressurized. Girardot constructed two fifteen-inch brass reflectors with one-inch glass ports and built-in micro-aqualungs.

We spent two months in a cold springtime sea photographing fixed flora and fauna. We worked in several deep wrecks to obtain data on their biological encrustation. Given a known date of sinking, scientists may study the wreck pictures for evidence on the rate of growth of the living blanket, something which could not be learned from undated rocks and reefs, where the organisms are sometimes six feet thick.

The temperature was fifty-two degrees Fahrenheit. It numbered our fingers and some of our wits. During one of the early dives, Ertaud forgot to release the valve that admitted compressed air to his reflector. Down in fourfold pressure, he was positioning his light when the glass port imploded with a crash that shook us up. One moment Ertaud was poised along the reef, the next he was falling like a man sprung through a gallows trap. The reflector which weighed nothing when ballasted with air, weighed thirty-five pounds empty. The brass bowl ripped the cable out of the camera, and Ertaud dropped to the bottom at an ear-punishing speed.

We swam down and regarded Ertaud's expression. It mingled the green drunken stare of the deep diver with an earthly remorse. He was vainly trying to raise the heavy bowl, which had cost us fifteen hundred dollars. Dumas straddled it and turned it face down. He lay on the floor, lifted one edge of the bowl and exhaled his bubbles under it. He had soon blown the bowl full of air. As an impromptu diving bell, the reflector lifted easily for a journey to the surface. In the bitter cold water, that same accident happened three times.

To add to our knowledge of light in the sea, I wanted to dive at night. I don't believe an undressed diver could honestly say that he was brave before a night dive. ("*I will have no man in my boat,*" says Starbuck in *Moby Dick*, "*who is not afraid of a whale.*") Some helmet divers are inured to working at night, because of their daytime familiarity with almost total darkness in dirty harbors and rivers, but I was scared of the sea at night.

I chose a ground I knew well, a rocky bottom twenty-five feet deep. It was a clear moonless summer night filled with bright stars. In the water billions of phosphorescent organisms vied with the stars and, as I put my mask under, the *noctiluca* redoubled their brightness, flashing on and off, like fireflies. Overhead the hull of the launch was a trembling oval of silver.

I revolved slowly down into the submerged Milky Way. I came upon rocks, bumps of ugly reality, and the dream crumbled. I could distinguish faint rock shapes in a narrow orbit. My imagination ranged into the black beyond, to the unseen recesses where night hunters, such as the conger and the moray, writhed in the merciless chase for food. This notion, more than any conscious intent to do so, caused my flashlight to go on.

A blinding conical beam hung in the water, extinguishing the tiny lights in its path. The flashlight ray made a creamy circle on the rocks. The light had the effect of plunging outer space into profound darkness. I could no longer see rocks beyond the area struck by light. I felt that hidden creatures were watching me from behind. I spun around, flashing the beam in all directions. The maneuver succeeded only in making me the blinder and losing my sense of direction.

It took resolution to turn the light off. In total darkness I began to swim cautiously across the rocks, often turning my head to dispel anxieties. In a bit my eyes readjusted and morbid shapes reformed in the environs. A faint shape moved, threw off a luminous cloud and flashed away like a comet—some surprised fish, awakened by the intruder. More fish jumped and fled.

After a while I mastered the fear and even drew some comfort from the fact that I was not down at night in tropical waters, among sharks. Before I returned to the boat, I believe I was enjoying the experience. The dive produced no observations of value.

I made another night dive under a full moon. The white light filtered strongly into the rocks. The landscape was revealed as far as one could see in daylight. But what a difference in mood! The rocks were enlarged to otherworldly dimensions. I imagined spectral human shapes and faces on them. The *noctiluca* sparks were almost extinguished; few micro-organisms were powerful enough to glow in the moonbeams. Wherever I looked I saw no living creature apart from the feeble pinpoints exerting themselves against the lunar power. There were no fish in the sea. When the moon rises above the horizon fishermen know that the fish have deserted the sea.

Epilogue

W HY in the world do you want to go down into the sea? is a riddle we are often asked by practical people. George Mallory was asked why he wanted to climb Mt. Everest, and his answer serves for us, too. "Because it is there," he said. We are obsessed with the incredible realm of oceanic life waiting to be known. The mean level of habitation on land, the home of all animals and plants, is a thin tissue shorter than a man. The living room of the oceans, which average twelve thousand feet in depth, is more than a thousand times the volume of the land habitat.

I have recounted how the first goggles led us underwater in simple and irresistible curiosity, and how that impulse entangled us in diving physiology and engineering, which produced the compressed-air lung. Our dives are now animated by the challenge of oceanography. We have tried to find the entrance to the great hydrosphere because we feel that the sea age is soon to come.

Since ancient times lonely men have tried to penetrate the sea. Sir Robert H. Davis has found records in each flourishing age of men who tried to make underwater breathing apparatus, most of them on swimming or free-walking principles. There are Assyrian bas-reliefs of men attempting impossible submersions while sucking on goatskin bellows. Leonardo da Vinci doodled several impractical ideas for diving lungs. Fevered Elizabethan craftsmen tinkered with leathern suits for diving. They failed because there was no popular economic movement to explore the sea, such as there was on land when Stephenson built the steam locomotive or when the Wrights took to the air.

Obviously man has to enter the sea. There is no choice in the

matter. The human population is increasing so rapidly and land resources are being depleted at such a rate, that we must take sustenance from the great cornucopia. The flesh and vegetables of the sea are vital. The necessity for taking mineral and chemical resources from the sea is plainly indicated by the intense political and economic struggles over tidal oil fields and the "continental shelf," by no means confined to Texas and California.

Our best independent diving range is only halfway down to the border of the shelf. We are not yet able to occupy the ground claimed by the statesmen. When research centers and industrialists apply themselves to the problem we will advance to the six-hundred-foot "dropoff" line. It will require much better equipment than the aqualung. The lung is primitive and unworthy of contemporary levels of science. We believe, however, that the conquerors of the shelf will have to get wet.